Principles and Standards for School Mathematics Navigations Series

Navigating through Number and Operations in Grades 9–12

Maurice J. Burke
Paul E. Kehle
Paul A. Kennedy
Dennis St. John

Maurice J. Burke
Grades 9–12 Editor

Peggy A. House
Navigations Series Editor

NATIONAL COUNCIL OF
TEACHERS OF MATHEMATICS

Copyright © 2006 by
The National Council of Teachers of Mathematics, Inc.
1906 Association Drive, Reston, VA 20191-1502
(703) 620-9840; (800) 235-7566; www.nctm.org

All rights reserved

Library of Congress Cataloging-in-Publication Data

Navigating through number and operations in grades 9-12 / Maurice J. Burke ... [et al.].
   p. cm. — (Principles and standards for school mathematics navigations series)
 Includes bibliographical references.
 ISBN 0-87353-585-5
 1. Mathematics—Study and teaching (Secondary)—Activity programs. 2. Numeration—Study and teaching (Secondary) 3. Numerical calculations—Study and teaching (Secondary) 4. Problem solving—Study and teaching (Secondary) 5. Reasoning—Study and teaching (Secondary) I. Burke, Maurice Joseph. II. Series.
 QA39.3.N38 2005
 510.71'2—dc22

                    2005027129

The National Council of Teachers of Mathematics is a public voice of mathematics education, providing vision, leadership, and professional development to support teachers in ensuring mathematics learning of the highest quality for all students.

   Permission to photocopy limited material from *Navigating through Number and Operations in Grades 9–12* (ISBN 0-87353-585-5) is granted for educational purposes. On the CD-ROM, the blackline masters may be downloaded and reproduced for classroom distribution; the applets may be used for instructional purposes in one classroom at a time. For permission to photocopy material or use it electronically for all other purposes, please access www.copyright.com or contact the Copyright Clearance Center, Inc. (CCC), 222 Rosewood Drive, Danvers, MA 01923, 978-750-8400. CCC is a not-for-profit organization that provides licenses and registration for a variety of users. Permission does not automatically extend to any items identified as reprinted by permission of other publishers and copyright holders. Such items must be excluded unless separate permissions are obtained. It will be the responsibility of the user to identify such materials and obtain the permissions.

   The publications of the National Council of Teachers of Mathematics present a variety of viewpoints. The views expressed or implied in this publication, unless otherwise noted, should not be interpreted as official positions of the Council.

Printed in the United States of America

NAVIGATIONS SERIES

GRADES 9–12

Table of Contents

About This Book vii

Introduction 1

Chapter 1
The Real Numbers 13
  When Fractions Are Whole 15
  Designing a Line 18
  Trigonometric Target Practice 23

Chapter 2
Number Theory and Algebra 27
  Counting Primes 29
  Complex Numbers and Matrices 33
  Solve That Number 38

Chapter 3
Number and Operations in the World 41
  Frequency, Scales, and Guitars 43
  Rock Around the Clock 52

Chapter 4
Extending Number and Operation Activities 61
  Number Triangles 63
  Perfect Squares 67
  Flooding a Water World 74

Looking Back and Looking Ahead 79

Appendix
Blackline Masters and Solutions 81
  When Fractions Are Whole 82
  Designing a Line 86
  Trigonometric Target Practice 89
  Counting Primes 91
  Adding Complex Numbers 95
  Multiplying Complex Numbers 98
  Solving Real Numbers 101
  Solving Complex Numbers 103
  Frequency, Scales, and Guitars 106
  Like Clockwork 111
  Encryption à la Mod 114
  Ciphering in Mod 31 118
  Make a Code / Break a Code 120
  Probing the Pattern 124
  It All Adds Up 127

iii

  Take a Trip around a Triangle 129
  Fair and Square 132
  Last-Digit Revelations 135
  The Singles Club 139
  Keeping It Legal 141
  Other Realms, Other Regions 146

  Solutions for the Blackline Masters 149

REFERENCES 170

Contents of the CD-ROM

Introduction

Table of Standards and Expectations, Number and Operations, Pre-K–12

Applets
Sound Wave
Driving with One Vector
Flying with Two Vectors

Blackline Masters
(All those listed above)

Readings from Publications of the National Council of Teachers of Mathematics

From Exploration to Generalization: An Introduction to Necessary and Sufficient Conditions
 Martin V. Bonsangue and Gerald E. Gannon
 Mathematics Teacher

So That's Why 22/7 Is Used for π!
 Maurice J. Burke and Diana L. Taggart
 Mathematics Teacher

Star Numbers and Constellations
 Richard L. Francis
 Mathematics Teacher

Calculator Cryptography
 Matthew Hall
 Mathematics Teacher

Roots in Music
 Don Houser
 Mathematics Teacher

Recursion and the Central Polygonal Numbers
 William A. Miller
 Mathematics Teacher

Counting Pizzas: A Discovery Lesson Using Combinatorics
    Gail Nord, Eric J. Malm, and John Nord
    Mathematics Teacher

Partitioning the Interior of a Circle with Chords
    Dennis Parker
    Mathematics Teacher

An Odd Sum
    Ray C. Shiflett and Harris S. Shultz
    Mathematics Teacher

Exploring Hill Ciphers with Graphing Calculators
    Dennis St. John
    Mathematics Teacher

About This Book

Navigating through Number and Operations in Grades 9–12 provides materials for you to use selectively in your classroom to deepen your students' understanding of number and operations. At the same time, working with these materials can consolidate your own awareness of the challenges of the Number and Operations Standard.

Principles and Standards for School Mathematics (National Council of Teachers of Mathematics [NCTM] 2000) places number and its nearly inseparable counterpart, operations, first among the Content Standards. This primacy of place recognizes the fundamental role that these entwined elements play in the development of all the other content strands of school mathematics. Number and operations certainly deserve this priority in grades 9–12. A large part of high school algebra is the study of operations on variables that stand for numbers, for example.

Principles and Standards is clear about the primary goal of instruction for number and operations in grades 9–12: "Whereas middle-grades students should have been introduced to irrational numbers, high school students should develop an understanding of the system of real numbers" (NCTM 2000, p. 291). Three words hold the keys to what high school students must attain over and above what they achieved in middle school: *understanding*, *system*, and *real*. An introduction to irrational numbers suffices for middle-grades students, but high school students must develop an understanding of the system of real numbers. A *system* is a unified set rather than a simple collection, and the real numbers are the fundamental numbers that students will need for measuring and counting in the world around them. In addition, as *Principles and Standards* emphasizes, students will need the system of real numbers to develop understandings of such new systems as vectors, matrices, and complex numbers.

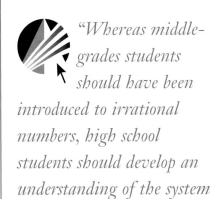

"Whereas middle-grades students should have been introduced to irrational numbers, high school students should develop an understanding of the system of real numbers."
(NCTM 2000, p. 291)

Overview of the Chapters

The book's four chapters present activities that show how to achieve goals of the Number and Operations Standard for grades 9–12. A consideration of the real numbers in chapter 1 prepares the way for an examination of number theory and algebra in chapter 2. Chapter 3 explores some applications of number and operations in the real world, and chapter 4 concludes the book by investigating some extensions of basic concepts in the world of mathematics.

Chapter 1, "The Real Numbers," focuses on the relationship between the rational numbers and the irrational numbers, the two systems that make up the real number system. From one perspective, these classes of numbers are extremely different. From another perspective, however, they are not so very unlike. Students encounter both perspectives in the chapter's three activities—When Fractions Are Whole, Designing a Line, and Trigonometric Target Practice:

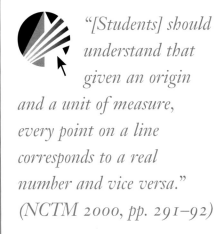

"[Students] should understand that given an origin and a unit of measure, every point on a line corresponds to a real number and vice versa."
(NCTM 2000, pp. 291–92)

See "So That's Why 22/7 Is Used for π!" (Burke and Taggart 2002; available on the CD-ROM) for a graphical method of building students' understanding of rational approximations for irrational numbers.

- When Fractions Are Whole helps students build and deepen their sense of rational and irrational numbers as fundamentally different. Students use what they know about simplifying fractions to discover—and prove—that they cannot express several common classes of numbers as rational numbers. They see, in other words, that these numbers are irrational.

- Designing a Line helps students take the opposite perspective. Now they consider rational and irrational numbers as not so very different from one another. *Principles and Standards* recommends that students understand the real number line—the geometric interpretation of real numbers that treats rational and irrational numbers in the same way—as points on a line. This activity focuses on geometric representations of both real numbers and the operations on real numbers.

- Trigonometric Target Practice pushes the students' understanding of rational and irrational numbers a step further, helping them realize that irrational numbers are more pervasive than they might have thought. The activity provides a graphical way of "seeing" when a number is irrational.

Principles and Standards also asserts that high school students "should understand that irrational numbers can only be approximated by fractions or by terminating or repeating decimals" (NCTM 2000, p. 292). Our inability to represent irrational numbers exactly by these means reflects an important element of the relationship between rational and irrational numbers. Although the book does not treat this aspect explicitly, Burke and Taggart (2002) have investigated how teachers might approach this issue with high school students. Their work appears as supplemental reading on the CD-ROM that accompanies the book.

Chapter 2, "Number Theory and Algebra," illustrates ways to help students understand the properties of number and operation systems, as *Principles and Standards* recommends. Students investigate properties of systems and apply them in exploring new systems in the chapter's three activities—Counting Primes, Complex Numbers and Matrices, and Solve That Number:

- Counting Primes engages students in exploring the divisibility of whole numbers and proving results that are easy to believe from considerations of the number line. The students then use these results to prove that the set of prime numbers is infinite.

- Complex Numbers and Matrices shows students how they can represent the complex number system by systems of vectors and matrices based on the real numbers. The activity engages the students in a cycle of exploring and justifying as they compare the complex number system with the system of matrices that they encounter in the activity. Their investigation of properties shows them that this system of matrices is isomorphic to the complex number system.

- Solve That Number sheds light on the distinction between algebraic and transcendental numbers. An excursion into the theory of equations adds a new twist to the algebra that students have previously encountered. Instead of receiving equations and finding numbers

"In grades 9–12 all students should …
- *compare and contrast the properties of numbers and number systems … ;*
- *understand vectors and matrices as systems that have some of the properties of the real-number system;*
- *use number-theory arguments to justify relationships involving whole numbers."*

(NCTM 2000, p. 290)

that solve them, the students now receive numbers and find polynomial equations, with integer coefficients, that each number solves. The activity refers to this process informally as "solving the number." In the course of the activity, the students encounter numbers that they cannot "solve," such as π. Through such experiences, they gain an appreciation of the difference between algebraic and transcendental numbers. This new insight, in turn, helps them recognize how prevalent transcendental numbers are and what mathematicians mean when they refer to the trigonometric functions as transcendental functions.

Chapter 3, "Numbers and Operations in the World," shows how to help students find and apply ideas about number and operations in the real world. The everyday importance of these concepts is clear in the chapter's two activities—Frequency, Scales, and Guitars, and Rock Around the Clock:

- Frequency, Scales, and Guitars illustrates the important role that irrational numbers play in the world of music. The activity demonstrates the significance of the twelfth root of two in the design and tuning of guitars and other musical instruments. High school students learn that real numbers in their world are often extremely large or extremely small, and they are often irrational. An applet activity, Sound Wave, adapted from the Illuminations Web site, appears on the accompanying CD-ROM as a supplement to this exploration.

- Rock Around the Clock helps students learn another important fact as they discover that the operations they need to work with numbers in the world around them are not limited to the "big four"—addition, subtraction, multiplication, and division. The students use multiplication of integers modulo n to make sense of a system of message-coding schemes known as *position ciphers*. A firm understanding of the properties of the basic operations ($+$, $-$, $\times$, $\div$) on integers, rational numbers, and real numbers allows students to extend their insights to new operations, and these discoveries, in turn, empower them to grasp mathematics in the world around them.

Chapter 4, "Extending Number and Operation Activities," the book's final chapter, shows how seemingly simple activities focusing on number and operation can lead students to mathematical investigations of increasing depth and sophistication. Such rewards are characteristic of any curriculum that emphasizes the five major aspects of "doing" mathematics—problem solving, reasoning and proof, communication, connections, and representation. These are the facets of mathematical study that *Principles and Standards* highlights in the Process Standards. These processes play central roles in this final chapter's three activities —Number Triangles, Perfect Squares, and Flooding a Water World:

- Number Triangles begins simply, with a problem-solving activity similar to ones that students encounter frequently at the elementary level. The students examine the pattern of numbers in a "number triangle" to supply the numbers that are missing in a second triangle. However, by reflecting on their solutions and solution processes,

Houser (2002; available on the CD-ROM) presents additional information on the use of the twelfth root of two in the design and tuning of some musical instruments.

Hall (2003) and St. John (1998) show how students can use matrices to extend their newly acquired ideas about position ciphers to other encryption methods. Both authors' pieces are available as supplemental reading on the accompanying CD-ROM.

Principles and Standards urges teachers to develop students' abilities to reason with and prove mathematical propositions in whole number contexts such as that in the activity Perfect Squares. Bosangue and Gannon (2003), Francis (1993), Miller (1991), and Shiflett and Shultz (2002) all discuss similar investigations and ideas for helping students delve deeper. Their work appears on the CD-ROM.

the students move to a consideration of such higher-level issues as solvability and the effectiveness of algorithms.

- Perfect Squares opens with an exploration of the properties of the class of natural numbers known as the perfect squares. The investigation quickly directs students' attention to the last digits in very large perfect squares—numbers that conventional calculators typically cannot handle. This consideration in turn leads to an investigation of a number system consisting of the last digits of integers under multiplication—a system that is equivalent to the integers modulo 10.

- Flooding a Water World allows students to explore simple networks of towers and dikes erected to create habitable regions in an imaginary world covered by water. The students establish relationships among the numbers of towers, dikes, and regions in a "legal" network. By working in this accessible, concrete context, the students easily discover and then apply several powerful ideas from graph theory and geometry. This activity is one of many counting tasks that can lead high school students to rich mathematical experiences without using permutations or combinations.

Using the Book

Needless to say, the Standards-based activities in this book are merely snapshots of possible activities, and they cannot do justice to all the recommendations in the Number and Operations Standard. For example, the book does not include an activity that delves into counting techniques for permutations and combinations. This significant topic and others of equal importance, including vectors and matrices, receive only the briefest of nods here. To provide some ideas of possible approaches to vectors, the accompanying CD-ROM includes two applet activities, Driving with One Vector and Flying with Two Vectors, adapted from electronic examples at NCTM's Web site.

"Counting Pizzas: A Discovery Lesson Using Combinatorics" (Nord et al. 2002; available on the CD-ROM) is a useful article on teaching combinatoric counting techniques.

In addition, it is very important to emphasize that for nearly every major recommendation in *Principles and Standards*, the journals of the National Council of Teachers of Mathematics include many useful articles. The topic of permutations and combinations, for example, offers a wealth of material from which to choose. The CD-ROM includes one of these pieces, and the activity Flooding a Water World in chapter 4 shows that rich and meaningful counting activities at the high school level need not involve combinations and permutations.

Navigating through the book and the CD-ROM

The appendix of *Navigating through Number and Operations in Grade 9–12* presents activity sheets for students as reproducible blackline masters. An icon in the margin (see the key on p. xi) signals all the blackline pages. You can make copies of these pages from the book or print them directly from the accompanying CD-ROM. Solutions to the problems posed in the blackline masters appear in the appendix and on the CD.

As this preface has already indicated, the CD-ROM also includes a number of pieces for teachers' professional development. A second icon in the text identifies all supplemental materials on the CD-ROM, including the book's three applet activities—Sound Wave, Driving with One Vector, and Flying with Two Vectors.

Throughout the book, margin notes supply teaching tips as well as pertinent statements from *Principles and Standards for School Mathematics*. A third icon flags these quotations, which highlight the fundamental notion that students should master the processes of mathematics and see mathematics as an integrated whole.

It is the highest hope of the authors and editors that the ideas and activities in this book are clear and useful. If the book encourages reflection on the NCTM Standards and the support that they can offer in the classroom, it will have served its primary purpose.

Key to Icons

Principles and Standards

CD-ROM

Blackline Master

Three different icons appear in the book, as shown in the key. One alerts readers to material quoted from *Principles and Standards for School Mathematics,* another points them to supplementary materials on the CD-ROM that accompanies the book, and a third signals the blackline masters and indicates their locations in the appendix.

Grades 9–12

Navigating through Number and Operations

Introduction

What could be more fundamental in mathematics than numbers and the operations that we perform with them? Thus, it is no surprise that Number and Operations heads the list of the five Content Standards in *Principles and Standards for School Mathematics* (NCTM 2000). Yet, numbers and arithmetic are so familiar to most of us that we run the risk of underestimating the deep, rich knowledge and proficiency that this Standard encompasses.

Fundamentals of an Understanding of Number and Operations

In elaborating the Number and Operations Standard, *Principles and Standards* recommends that instructional programs from prekindergarten through grade 12 enable all students to—

- understand numbers, ways of representing numbers, relationships among numbers, and number systems;
- understand meanings of operations and how they relate to one another;
- compute fluently and make reasonable estimates.

The vision that *Principles and Standards* outlines in the description of this Standard gives Number and Operations centrality across the entire mathematics curriculum. The *Navigating through Number and Operations* volumes flesh out that vision and make it concrete in activities for students in four grade bands: prekindergarten through grade 2, grades 3–5, grades 6–8, and grades 9–12.

Understanding numbers, ways of representing numbers, relationships among numbers, and number systems

Young children begin to develop primitive ideas of number even before they enter school, and they arrive in the classroom with a range of informal understanding. They have probably learned to extend the appropriate number of fingers when someone asks, "How old are you?" and their vocabulary almost certainly includes some number words. They are likely to be able to associate these words correctly with small collections of objects, and they probably have been encouraged to count things, although they may not yet have mastered the essential one-to-one matching of objects to number names. During the years from prekindergarten through grade 2, their concepts and skills related to numbers and numeration, counting, representing and comparing quantities, and the operations of adding and subtracting will grow enormously as these ideas become the focus of the mathematics curriculum.

The most important accomplishments of the primary years include the refinement of children's understanding of counting and their initial development of number sense. Multiple classroom contexts offer numerous opportunities for students to count a myriad of things, from how many children are in their reading group, to how many cartons of milk their class needs for lunch, to how many steps they must take from the chalkboard to the classroom door. With experience, they learn to establish a one-to-one matching of objects counted with number words or numerals, and in time they recognize that the last number named is also the total number of objects in the collection. They also discover that the result of the counting process is not affected by the order in which they enumerate the objects. Eventually, they learn to count by twos or fives or tens or other forms of "skip counting," which requires that quantities be grouped in certain ways.

Though children initially encounter numbers by counting collections of physical objects, they go on to develop number concepts and the ability to think about numbers without needing the actual objects before them. They realize, for example, that five is one more than four and six is one more than five, and that, in general, the next counting number is one more than the number just named, whether or not actual objects are present for them to count. Through repeated experience, they also discover some important relationships, such as the connection between a number and its double, and they explore multiple ways of representing numbers, such as modeling six as six ones, or two threes, or three twos, or one more than five, or two plus four.

Young children are capable of developing number concepts that are more sophisticated than adults sometimes expect. Consider the prekindergarten child who explained her discovery that some numbers, like 2 and 4 and 6, are "fair numbers," or "sharing numbers," because she could divide these numbers of cookies equally with a friend, but numbers like 3 or 5 or 7 are not "fair numbers," because they do not have this property.

As children work with numbers, they discover ways of thinking about the relationships among them. They learn to compare two numbers to determine which is greater. If they are comparing 17 and 20, for example, they might match objects in two collections to see that 3 objects are "left over" in the set of 20 after they have "used up" the set of 17, or they might count on from 17 and find that they have to count three more numbers to get to 20. By exploring "How many more?" and "How many less?" young children lay the foundations for addition and subtraction.

Continual work with numbers in the primary grades contributes to students' development of an essential, firm understanding of place-value concepts and the base-ten numeration system. This understanding often emerges from work with concrete models, such as base-ten blocks or linking cubes, which engage students in the process of grouping and ungrouping units and tens. They must also learn to interpret, explain, and model the meaning of two- and three-digit numbers written symbolically. By the end of second grade, *Principles and Standards* expects students to be able to count into the hundreds, discover patterns in the numeration system related to place value, and compose (create through different combinations) and decompose (break apart in different ways) two- and three-digit numbers.

In addition, students in grade 2 should begin to extend their understanding of whole numbers to include early ideas about fractions. Initial experiences with fractions should introduce simple concepts, such as the idea that halves or fourths signify divisions of things into two or four equal parts, respectively.

As students move into grades 3–5, their study of numbers expands to include larger whole numbers as well as fractions, decimals, and negative numbers. Now the emphasis shifts from addition and subtraction to multiplication and division, and the study of numbers focuses more directly on the multiplicative structure of the base-ten numeration system. Students should understand a number like 435 as representing $(4 \times 100) + (3 \times 10) + (5 \times 1)$, and they should explore what happens to numbers when they multiply or divide them by powers of 10.

The number line now becomes an important model for representing the positions of numbers in relation to benchmarks like 1/2, 1, 10, 100, 500, and so on. It also provides a useful tool at this stage for representing fractions, decimals, and negative integers as well as whole numbers.

Concepts of fractions that the curriculum treated informally in the primary grades gain new meaning in grades 3–5 as students learn to interpret fractions both as parts of a whole and as divisions of numbers. Various models contribute to students' developing understanding. For example, an area model in which a circle or a rectangle is divided into equal parts, some of which are shaded, helps students visualize fractions as parts of a unit whole or determine equivalent fractions.

Number-line models are again helpful, allowing students to compare fractions to useful benchmarks. For instance, they can decide that 3/5 is greater than 1/3 because 3/5 is more than 1/2 but 1/3 is less than 1/2,

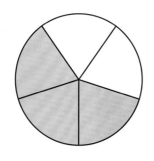

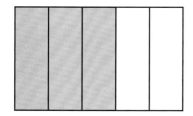

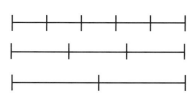

or they can recognize that 9/10 is greater than 7/8 because 9/10 is closer to 1 than 7/8 is. Parallel number lines, such as one marked in multiples of 1/3 and another in multiples of 1/6, can help students identify equivalences.

During these upper elementary years, students also encounter the concept of percent as another model for a part of a whole. Their work should help them begin to develop benchmarks for common percentages, such as 25 percent, $33\frac{1}{3}$ percent, or 50 percent.

In grades 6–8, students expand their understanding of numbers to include the integers, and now they learn how to add, subtract, multiply, and divide with negative as well as positive numbers. Developing a deeper understanding of rational numbers is another very important goal for these students, who must increase their facility in working with rational numbers represented by fractions, decimals, and percents.

At this level, the curriculum places particular emphasis on developing proportional reasoning, which requires students to understand and use ratios, proportions, and rates to model and solve problems. Fraction strips, circles, number lines, area models, hundreds grids, and other physical models provide concrete representations from which students can draw conceptual meaning as they hone their understanding of rational numbers. Exposure to these models develops students' abilities to translate fluently from one representation to another, to compare and order rational numbers, and to attach meaning to rational numbers expressed in different but equivalent forms.

The concept of proportionality, which is a central component of the middle-school curriculum, serves to connect many aspects of mathematics, such as the slope of the linear function $y = mx$ in algebra, the scale factor in measurements on maps or scale drawings, the ratio of the circumference to the diameter of a circle (π) in geometry, or the relative frequency of a statistic in a set of data. Thus, students have numerous opportunities to develop and use number concepts in multiple contexts and applications. In some of those contexts, students encounter very large or very small numbers, which necessitate scientific notation and a sense of orders of magnitude of numbers.

Finally, students in grades 6–8 are able to focus more directly on properties of numbers than they were at earlier stages of development. They can investigate such key ideas as the notions of factor and multiple, prime and composite numbers, factor trees, divisibility tests, special sets (like the triangular and square numbers), and many interesting number patterns and relationships, including an introduction to some irrational numbers, such as $\sqrt{2}$.

When students move on to grades 9–12, their understanding of number should continue to grow and mature. In these grades, students customarily encounter many problems, both in mathematics and in related disciplines like science or economics, where very large and very small numbers are commonplace. In working such problems, students can use technology that displays large and small numbers in several ways, such as 1.219 E17 for 1.219 (10^{17}), and they need to become fluent in expressing and interpreting such quantities.

High school students also have many opportunities to work with irrational numbers, and these experiences should lead them to an understanding of the real number system—and, beyond that, to an understanding of number systems themselves. Moreover, students in grades 9–12 should develop an awareness of the relationship of those systems to various types of equations. For example, they should understand that the equation $A + 5 = 10$ has a whole-number solution, but the equation $A + 10 = 5$ does not, though it does have an integer solution. They should recognize that the equation $10 \cdot A = 5$ requires the rational numbers for its solution, and the equation $A^2 = 5$ has a real-number solution, but the equation $A^2 + 10 = 5$ is solved in the complex numbers.

Students should also understand the one-to-one correspondence between real numbers and points on the number line. They should recognize important properties of real numbers, such as that between any two real numbers there is always another real number, or that irrational numbers can be only approximated, but never represented exactly, by fractions or repeating decimals.

In grades 9–12, students also encounter new systems, such as vectors and matrices, which they should explore and compare to the more familiar number systems. Such study will involve them in explicit examination of the associative, commutative, and distributive properties and will expand their horizons to include a system (matrices) in which multiplication is not commutative. Using matrices, students can represent and solve a variety of problems in other areas of mathematics. They can find solutions to systems of linear equations, for instance, or describe a transformation of a geometric figure in the plane. Using algebraic symbols and reasoning, students also can explore interesting number properties and relationships, determining, for example, that the sum of two consecutive triangular numbers is always a square number and that the sum of the first N consecutive odd integers is equal to N^2.

Understanding meanings of operations and how operations relate to one another

As young children in prekindergarten through grade 2 learn to count and develop number sense, they simultaneously build their understanding of addition and subtraction. This occurs naturally as children compare numbers to see who collected more stickers or as they solve problems like the following: "When Tim and his dad went fishing, they caught seven fish. Tim caught four of the fish. How many did his dad catch?" Often, children use concrete materials, such as cubes or chips, to model "joining" or "take-away" problems, and they develop "counting on" or "counting back" strategies to solve problems about "how many altogether?" and "how many more?" and similar relationships.

Even at this early stage, teachers who present problems in everyday contexts can represent the problem symbolically. For example, teachers can represent the problem, "How many more books does Emily need to read if she has already read 13 books and wants to read 20 books before the end of the school year?" as $13 + \square = 20$ or as $20 - \square = 13$. Such expressions help students to see the relationship between addition and subtraction.

Young children also build an understanding of the operations when they explain the thinking behind their solutions. For example, a child who had just celebrated his sixth birthday wondered, "How much is

6 and 7?" After thinking about the problem for a moment, he decided that 6 + 7 = 13, and then he explained how he knew: "Well, I just had a birthday, and for my birthday I got two 'five dollars,' and my $5 and $5 are $10, so 6 and 6 should be 12, and then 6 and 7 must be 13."

As young students work with addition and subtraction, they should also be introduced to the associative and commutative properties of the operations. They should learn that when they are doing addition, they can use the numbers in any order, but they should discover that this fact is not true for subtraction. Further, they should use the commutative property to develop effective strategies for computation. For example, they might rearrange the problem 3 + 5 + 7 to 3 + 7 + 5 = (3 + 7) + 5 = 10 + 5 = 15.

Early work with addition and subtraction also lays the conceptual groundwork for later study of operations. Multiplication and division are all but evident when students repeatedly add the same number—for example, in skip-counting by twos or fives—or when they solve problems requiring that a collection of objects be shared equally among several friends. The strategies that young children use to solve such problems, either repeatedly adding the same number or partitioning a set into equal-sized subsets, later mature into computational strategies for multiplication and division.

The operations of multiplication and division, and the relationships between them, receive particular emphasis in grades 3–5. Diagrams, pictures, and concrete manipulatives play important roles as students deepen their understanding of these operations and develop their facility in performing them.

For example, if an area model calls for students to arrange 18 square tiles into as many different rectangles as they can, the students can relate the three possible solutions (1 by 18, 2 by 9, and 3 by 6) to the factors of 18. Similar problems will show that some numbers, like 36 or 64, have many possible rectangular arrangements and hence many factors, while other numbers, like 37 or 41, yield only one solution and thus have only two factors. By comparing pairs of rectangular arrangements, such as 3 by 6 and 6 by 3, students can explore the commutative property for multiplication. As illustrated in the three examples below,

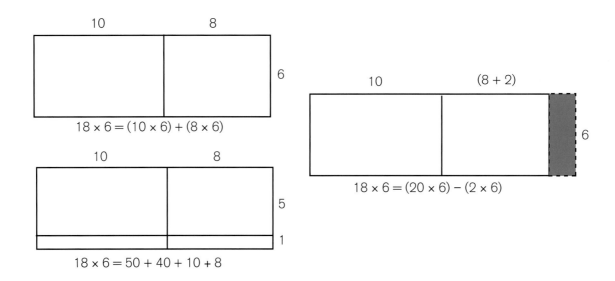

by decomposing an 18-by-6 area model, students can develop an understanding of the distributive property.

Other models for multiplication might involve rates or combinations. In grades 3–5, a typical problem involving a rate might be, "If 4 pencils cost 69¢, how much will a dozen pencils cost?" Problems involving combinations at this level are often similar to the following: "How many different kinds of meat-and-cheese sandwiches can we make if we have 2 kinds of bread (white and wheat), 4 kinds of meat (beef, ham, chicken, and turkey), and 3 kinds of cheese (Swiss, American, and provolone)?" (See the tree diagram below.)

To develop students' understanding of division, teachers should engage them in working with two different models—a partitioning model ("If you have 36 marbles and want to share them equally among 4 people, how many marbles should each person receive?") and a repeated-subtraction model ("If you have 36 marbles and need to place 4 marbles into each cup in a game, how many cups will you fill?"). Students should be able to represent both models with manipulatives and diagrams.

In exploring division, students in grades 3–5 will inevitably encounter situations that produce a remainder, and they should examine what the remainder means, how large it can be for a given divisor, and how to interpret it in different contexts. For example, arithmetically,

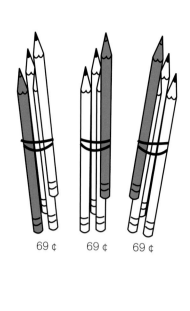

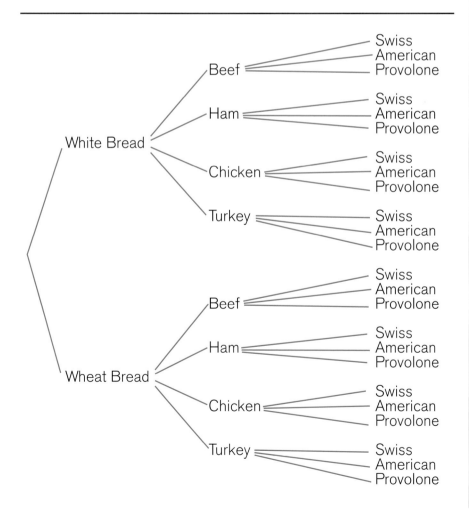

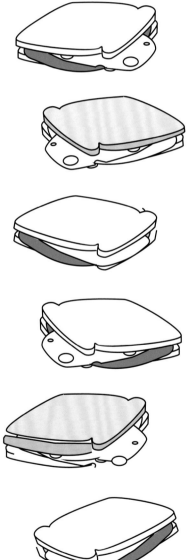

$28 \div 5 = 5\frac{3}{5}$, but consider the solutions to each of the following problems:

- "Compact disks are on sale for $28 for 5 disks. How much should one disk cost?" ($5.60)
- "Muffins are packaged 5 to a box for the bake sale. How many boxes can you make up if you bake 28 muffins?" (5).
- "Parents will be transporting children in minivans for the class field trip. Each van can take 5 children. The class has 28 children. How many vans will parents need to drive for the trip?" (6).

The understanding of all four operations that students build with whole numbers in the upper elementary grades broadens during grades 6–8, when they apply those operations to fractions, decimals, percents, and integers. Moreover, as students operate with rational numbers and integers, they encounter new contexts that may challenge their conceptual foundations. For example, when students are multiplying or dividing with fractions or decimals between 0 and 1, they see results that expose as misconceptions the commonly held beliefs that "multiplication makes larger" and "division makes smaller."

Other challenges that middle-grades students must confront include understanding when the result of a computation with integers is positive and when it is negative, knowing how to align decimals in computations with decimal fractions, and recognizing where in an answer to place a decimal point. Operating with fractions has proven difficult for many students. Lacking conceptual understanding, many have tried to get by with rote application of procedures that they don't understand. In the middle grades, therefore, it is important that students develop an understanding of the meaning of such concepts as numerator, denominator, and equivalent fractions and their roles in adding and subtracting fractions.

Middle school students need to model and compare expressions that are frequently subject to confusion, such as "divide by 2," "multiply by 1/2," and "divide by 1/2," and they must see that different models of division are sometimes required to give meaning to such ideas. For example, "divide by 2" can be modeled by a partitioning model ("separate the amount into two equal quantities"), but "divide by 1/2" is more appropriately represented by a repeated-subtraction model:

"You made $2\frac{3}{4}$ gallons of lemonade. How many $\frac{1}{2}$-gallon bottles can you fill?"

$$\left(2\frac{3}{4} \div \frac{1}{2} = 5, \text{ with a remainder of } \frac{1}{4} \text{ gallon}\right)$$

Encouraging students to estimate and evaluate the reasonableness of the results of their computations is important in helping them expand their number sense.

As students' algebraic concepts grow during grades 6–8, they will also frequently face computations involving variables, and they will need to extend their understanding of the operations and their properties to encompass simplification of and operations with algebraic expressions. Understanding the inverse relationship between addition and subtraction, between multiplication and division, and between "square" and "square root" will be important in such tasks.

In grades 9–12, students should go beyond producing the results of specific computations to generalize about operations and their properties and to relate them to functions and their graphs. For example, they should describe and compare the behavior of functions such as $f(x) = 2x$, $g(x) = x + 2$, $h(x) = x^2$, or $j(x) = \sqrt{x}$. They should reason about number relations, describing, for instance, the value of $a \cdot b$ where a and b are positive numbers and $a + b = 50$. They should understand and correctly apply the results of operating with positive or negative numbers when they are working with both equations and inequalities.

In addition, high school students should learn to perform operations in other systems. They should find vector sums in the plane, add and multiply matrices, or use multiplicative reasoning to represent counting problems and combinatorics.

Computing fluently and making reasonable estimates

Although an understanding of numbers and the meanings of the various operations is essential, it is insufficient unless it is accompanied by the development of computational proficiency and a sense of the reasonableness of computational results. Computational skills emerge in the prekindergarten and early elementary years in conjunction with students' developing understanding of whole numbers and counting.

Young children's earliest computational strategies usually involve counting. As they think about number problems involving addition or subtraction, young students devise different solution schemes, and teachers should listen carefully to their students' explanations of these thinking strategies. Encouraging children to explain their methods and discussing different students' strategies with the class helps students deepen their understanding of numbers and operations and refine their computational abilities.

At first, young children rely heavily on physical objects to represent numerical situations and relationships, and they use such objects to model their addition and subtraction results. Over time, they learn to represent the same problems symbolically, and eventually they carry out the computations mentally or with paper and pencil, without needing the actual physical objects. Students should have enough experience and practice to master the basic one-digit addition and subtraction combinations, and they should combine that knowledge with their understanding of base-ten numeration so that, by the end of grade 2, they can add and subtract with two-digit numbers.

As students become more proficient with addition and subtraction, teachers can help them examine the efficiency and generalizability of their invented strategies and can lead them to an understanding of standard computational algorithms. When students understand the procedures that they are employing, they are able to carry them out with accuracy and efficiency.

In grades 3–5, students should extend their knowledge of basic number combinations to include single-digit multiplication and division facts, and by the end of the upper elementary years they should be able to compute fluently with whole numbers. As students develop their computational proficiency, teachers should guide them in examining and explaining their various approaches and in understanding algorithms for addition, subtraction, multiplication, and division and employing them effectively. In turn, teachers must understand that there is more than one algorithm for each of the operations, and they should recognize that the algorithms that are meaningful to students may not be the ones that have traditionally been taught or that some people have come to assume offer "the right way" to solve a problem.

In grades 3–5, students are beginning to work with larger numbers, and it is important for them to develop a strong sense of the reasonableness of a computational result and a facility in estimating results before computing. It will often be appropriate for students to use calculators when they are working with larger numbers. At other times, paper and pencil may be appropriate, or it may be reasonable for teachers to expect mental computation. Teachers and students should discuss various situations to assist students in developing good judgment about when to use mental arithmetic, paper and pencil, or technology for whole-number computation.

Other aspects of computational fluency in the 3–5 grade band involve understanding the associative, commutative, and distributive properties and seeing how those properties can be used to simplify a computation. Students at this level will also encounter problems that require the introduction of order-of-operations conventions.

While students in grades 3–5 are honing their skills with whole-number computation, they also will be spending a great deal of time developing an understanding of fractions and decimals. However, computing with rational numbers should not be the focus of their attention yet. Rather, students should apply their understanding of fractions and decimals and the properties of the operations to problems that include fractions or decimals. For example, "How many sheets of construction paper will Jackie need to make 16 Halloween decorations if each decoration uses $2\frac{1}{4}$ sheets of paper?" General procedures for calculating with rational numbers and integers will be the focus of instruction in the next grade band.

In grades 6–8, students learn methods for computing with fractions and decimals as extensions of their understanding of rational numbers and their facility in computing with whole numbers. As with whole-number computation, students develop an understanding of computing with fractions, decimals, and integers by considering problems in context, making estimates of reasonable expectations for the results, devising and explaining methods that make sense to them, and comparing their strategies with those of others as well as with standard algorithms. When calculating with fractions and decimals, students must learn to assess situations and decide whether an exact answer is required or whether an estimate is appropriate. They should also develop useful benchmarks to help them assess the reasonableness of results when they are calculating with rational numbers, integers, and percents. Computational fluency at the middle grades also includes a facility in reasoning about and solving problems involving proportions and rates.

In grades 9–12, students should extend their computational proficiency to real numbers and should confidently choose among mental mathematics, paper-and-pencil calculations, and computations with technology to obtain results that offer an appropriate degree of precision. They should perform complex calculations involving powers and roots, vectors, and matrices, as well as real numbers, and they should exhibit a well-developed number sense in judging the reasonableness of calculations, including calculations performed with the aid of technology.

Numbers and Operations in the Mathematics Curriculum

Without numbers and operations there would be no mathematics. Accordingly, the mathematics curriculum must foster the development of both number sense and computational fluency across the entire pre-K–12 continuum. The Number and Operations Standard describes the core of understanding and proficiency that students are expected to attain, and a curriculum that leads to the outcomes envisioned in this Standard must be coherent, developmental, focused, and well articulated across the grades. At all levels, students should develop a true understanding of numbers and operations that will undergird their development of computational fluency.

The *Navigating through Number and Operations* books provide insight into the ways in which the fundamental ideas of number and operations can develop over the pre-K–12 years. These Navigations volumes, however, do not—and cannot—undertake to describe a complete curriculum for number and operations. The concepts described in the Number and Operations Standard regularly apply in other mathematical contexts related to the Algebra, Geometry, Measurement, and Data Analysis and Probability Standards. Activities such as those described in the four *Navigating through Number and Operations* books reinforce and enhance understanding of the other mathematics strands, just as those other strands lend context and meaning to number sense and computation.

The development of mathematical literacy relies on deep understanding of numbers and operations as set forth in *Principles and Standards for School Mathematics*. These *Navigations* volumes are presented as a guide to help educators set a course for the successful implementation of this essential Standard.

NAVIGATIONS SERIES
GRADES 9–12

NAVIGATING through NUMBER and OPERATIONS

Chapter 1
The Real Numbers

Principles and Standards for School Mathematics (NCTM 2000) calls for instruction in grades 9–12 that will increase students' understanding of number, operation, and number systems. In the years before high school, students learned about whole numbers and integers as systems of numbers and operations, and they encountered rational and irrational numbers as the complementary sets that make up the real numbers. Nevertheless, in grades 9–12, they still must master many fundamental concepts related to these number systems. This chapter presents some of these ideas in activities that can give new depth and sophistication to high school students' understanding of the real number system. For instance, students usually enter high school knowing that $\sqrt{2}$ is irrational. But how many students can explain why this is true or what its significance is? Some students would say that *irrational* means that $\sqrt{2}$ cannot be expressed as a fraction. Yet, they might be confused to learn that

$$\sqrt{2} = \frac{2}{\sqrt{2}},$$

which is a fraction, though not a rational one. Moreover, even students who understand that $\sqrt{2}$ is not a rational fraction might have misgivings if someone asked them to think about an irrational number as an infinite, nonrepeating decimal.

In fact, the idea of an irrational number as an infinite, nonrepeating decimal may confound students who view arithmetic operations in terms of the usual algorithms for integer, decimal, and rational number operations. Students who customarily think of multiplication, for

"Whereas middle-grades students should have been introduced to irrational numbers, high school students should develop an understanding of the system of real numbers."
(NCTM 2000, p. 291)

example, in the conventional right-to-left decimal algorithm would be hard-pressed to explain how to multiply $\sqrt{2}$ and π as infinite decimals. They might not even be able to make sense of such a multiplication.

Nevertheless, as algebraic reasoning skills mature and an understanding of geometric relationships grows in grades 9–12, students should have opportunities to deepen their understanding of the real numbers and their representations. This chapter's activities provide several such opportunities.

The first activity, When Fractions Are Whole, investigates issues that underlie the distinction between *rational* and *irrational*. By paying close attention to the ideas involved in simplifying rational fractions, the students prove that the square root of a positive integer is not rational unless the positive integer is a perfect square. In other words, the students prove that if a positive integer is not a perfect square, then its square root is irrational.

The second activity, Designing a Line, presents the real number system from a geometrical point of view, highlighting one of the most important representations of number—the real number line. This representation overlooks the distinction between rational and irrational numbers and treats numbers of both classes in the same way—as points on a number line. It also overlooks distinctions among arithmetic operations as they pertain to rational and irrational numbers by treating addition and multiplication as geometric transformations of points on the number line.

The third activity, Trigonometric Target Practice, gives visual meaning to irrational numbers and illustrates how prevalent such numbers are. A line that passes through the origin and has an irrational slope never passes through a lattice point—a point whose coordinates are integers on a Cartesian grid. In this activity, students use graphing calculators to investigate graphs of lines of the form $y = \tan(\theta) \cdot x$, with the slope expressed as the tangent of the angle (in radians) that the line forms with the positive *x*-axis. In general, when θ is rational, the slope is irrational, and the line never passes through a lattice point other than (0, 0). The exception is the line formed when $\theta = 0$.

The activities in this chapter illustrate approaches to the study of real numbers from many different directions and representations. It is not sufficient for students to hear from an authority or read in their textbooks that every real number has a decimal representation and every decimal is a real number. The Number and Operation Standard calls for students to understand the real number system at a much deeper level than this. Thus, it is no accident that each of the following activities lets students encounter irrational numbers in connection with rational numbers and provides perspective on the meaning of *real number*.

When Fractions Are Whole

Goals

- Identify the conditions under which rational fractions can be equivalent to whole numbers
- Examine the implications of stating that a fraction is in simplest form
- Prove that $\sqrt{N}$ is irrational for any positive integer N unless N is a perfect square

Materials and Equipment

For each student—
- A copy of the activity sheet "When Fractions Are Whole"
- (Optional) A graphing calculator

Discussion

A positive rational fraction is a fraction $\frac{p}{q}$, where p and q are positive integers and $q \neq 0$. The activity sheet "When Fractions Are Whole" gives students this definition and tells them that $\frac{p}{q}$ is in simplest form when the greatest common divisor of p and q is 1. They consult a list of properties of rational fractions (see fig. 1.1) in steps 1–3.

pp. 82–85

1. The fraction is in simplest form.
2. The fraction is not in simplest form.
3. The fraction is in simplest form with denominator = 1.
4. The fraction is in simplest form with denominator ≠ 1.
5. The fraction is equivalent to a positive integer.
6. The fraction is not equivalent to a positive integer.
7. The fraction's square is equivalent to a positive integer.
8. The fraction's square is not equivalent to a positive integer.
9. The fraction's square is in simplest form.
10. The fraction's square is not in simplest form.
11. The fraction's denominator is a divisor of its numerator.
12. The fraction's numerator is a divisor of its denominator.

Fig. **1.1.**
Some properties of a rational fraction

Steps 1 and 2 direct the students to find fractions that exhibit specific combinations of properties on the list. You can have your students work in groups of three or four, checking one another's work and combining their results as they look for patterns. If the students do not think that any fraction exists that satisfies a specific combination of properties, they must justify this conclusion.

Step 3 asks students to reconsider the list of properties to complete two statements in such a way as to make them true:

Fact 1: If a fraction is in simplest form, then the square of the fraction _____.

Fact 2: If a fraction is in simplest form and is equivalent to a positive integer, then its denominator _____.

Before attempting to complete these statements, students should familiarize themselves thoroughly with the properties in the list, and they should agree within their groups that their examples in steps 1 and 2 are correct.

The objective of steps 1–3 is to help students discover that for fractions in the form $\frac{p}{q}$, where p and q are positive integers and $q \neq 0$, the following two statements are true:

Fact 1: If $\frac{p}{q}$ is in simplest form, then so is $\left(\frac{p}{q}\right)^2$, or $\frac{p^2}{q^2}$.

Fact 2: If $\frac{p}{q}$ is in simplest form and is equal to a positive integer, then $q = 1$.

The remainder of the activity poses a sequence of questions leading to a proof that the only positive integers that have rational square roots are the perfect squares. The students suppose that N is a positive integer and

$$\sqrt{N} = \frac{p}{q},$$

a rational number in simplest form. From this supposition, they can say

$$N = \frac{p^2}{q^2}.$$

But by referring to their completed statements in step 3, they can also say that since $\frac{p}{q}$ is in simplest form, $\frac{p^2}{q^2}$ is in simplest form as well (fact 1).

Moreover, since N is an integer, they can say that $q^2 = 1$ (fact 2). Therefore, they can conclude that $N = p^2$. Thus, if the square root of a positive integer N is rational, then N is a perfect square.

The contrapositive is of course also true: if a positive integer is not a perfect square, then its square root is not a rational number. In other words, its square root is irrational.

This argument is a direct proof of the fact that an infinite number of positive integers have irrational square roots. Traditional textbooks often show students that $\sqrt{2}$ is an irrational number through a proof by contradiction. Because the proof pivots on the idea of odd and even, students are unlikely to generalize it to the square roots of other integers.

In addition to demonstrating that the number of positive integers with irrational roots is infinite, a general proof has a second advantage. Students can extend its strategy to the cases of cube roots and higher-order roots. For example, they can easily replace fact 1 by the statement,

"[High school students] should understand the difference between rational and irrational numbers. Their understanding of irrational numbers needs to extend beyond π and $\sqrt{2}$."
(NCTM 2000, p. 292)

"If $\frac{p}{q}$ is in simplest form, then so is $\left(\frac{p}{q}\right)^3$, or $\frac{p^3}{q^3}$." Thus, if
$$\sqrt[3]{N} = \frac{p}{q},$$
the same line of reasoning that the students used in the case of square roots will lead them to the conclusion that $\sqrt[3]{N}$ is irrational for all integers N that are not perfect cubes.

Assessment

The students should fully understand facts 1 and 2 and accept them as true before going on in the activity. To ensure that all your students have grasped these ideas and are ready to proceed, you might take a break after they have completed steps 1–3 to discuss their responses. Otherwise, some students might be confused about where the argument is heading. Step 4 tells the students that N is a positive integer and $\sqrt{N}$ is equal is to $\frac{p}{q}$, a rational fraction in simplest form. The students then consider three questions:

a. What does $\left(\frac{p}{q}\right)^2$, or $\frac{p^2}{q^2}$, equal?

b. Is $\frac{p^2}{q^2}$ in simplest form?

c. What can they conclude about the denominator in $\frac{p^2}{q^2}$?

Consider introducing step 4 by telling your students that you are thinking of a positive integer N whose square root is $\frac{p}{q}$, a rational fraction in simplest form. Explain to them that their job is to narrow down the possible values of N so that they can try to guess the number you are thinking of.

For an overall assessment of students' mastery of the main ideas of this activity, ask them to prove that $\sqrt[3]{N}$ is irrational for all integers N that are not perfect cubes. Remind your students what it means to say that a positive integer is a perfect cube. If necessary, suggest to them that they modify fact 1 for the case of cubing a fraction.

Where to Go Next in Instruction

The activity When Fractions Are Whole can open students' eyes to the fact that many "familiar" real numbers are irrational. Indeed, the students prove that infinitely many real numbers are irrational. The activity highlights the distinction between a rational and an irrational number.

By contrast, the next activity, Designing a Line, helps students view the rational and irrational numbers as elements of equal status composing the real number system. The activity assists students in achieving this perspective by introducing them to a geometric way of representing the real numbers.

Designing a Line

Goals

- Identify irrational numbers and rational numbers as points on a number line
- Interpret addition and multiplication of rational and irrational numbers as operations on a number line
- Compare the infinite decimal interpretation of real numbers with the geometric interpretation with respect to the operations of addition and multiplication

Materials and Equipment

For each student—

- A copy of the activity sheet "Designing a Line"
- Two 3-by-5-inch index cards (unlined)
- A three-foot length of fabric ribbon, bias tape, or adding-machine paper tape for a number line. (Students must be able to write easily and neatly on the ribbon or tape. Fan-fold printer paper, cut into strips, will also work.)
- A pencil, pen, fine-tipped marker, or other appropriate device for marking on the ribbon or tape

For each group of four students—

- A sheet or two of regular binder paper (lined)
- (Optional) Two or three sheets of tagboard or construction paper and a roll of transparent tape
- (Optional for each pair of students) A compass and a straightedge

Discussion

In this activity, the students suppose that a fictitious archaeological association, the Ancient Egyptian Revival Society, wants them to recreate a system for measuring length from 2000 B.C. The students imagine that the society is mounting an exhibit and has asked them to make a number line based on a *palm*. For the task, each student has a strip of fabric or paper as well as a 3-by-5-inch index card to measure a palm (approximately 3 in.).

The activity sheet tells the students that the ancient Egyptians used a *palm* and a *cubit* as basic measures of length. A cubit was the distance from the elbow down the forearm to the tip of the middle finger. A palm was the width of the hand at the base of the four fingers. The Rhind Papyrus, one of the oldest extant mathematical documents, treats a palm as one-seventh of a cubit.

Students begin their work by placing marks on the number line to correspond with a variety of palm-lengths:

$$2, 3, 4, 5, 6, 7, \frac{1}{6}, \frac{2}{6}, \frac{3}{6}, \ldots, \frac{41}{6}, \sqrt{2}, \sqrt{3}, \sqrt{5}, \pi, 2\pi, \frac{\pi}{2}, \frac{3\pi}{2}.$$

pp. 86–88

Call your students' attention to the restriction that the society has imposed: in keeping with ancient technology and methods, the students must forgo using such tools as a ruler or a calculator. Instead, they must use geometry to devise methods of constructing fractional parts of a palm.

This goal will be challenging for many students, and some may come up with methods that quickly bring them back to the ruler. For instance, when constructing one-sixth of a palm, students might think of dividing 3 inches (the approximate size of a palm) by 6, thus determining that 1/6 of a palm is roughly 1/2 of an inch. These students might then want to use a ruler to measure 1/2 inch on their number lines. Tell them that this method would certainly work and has merit. However, the activity specifically asks them to use geometry.

The students can be quite inventive in the geometry that they choose to use. For example, to find sixths, a student can lay the three-inch side of the index card on lined notebook paper in such a way that it exactly spans six equal spaces between lines (see fig. 1.2). The intersections of the lines with the edge of the card will divide its three-inch edge into six equal parts.

Would the Ancient Egyptian Revival Society think that machine-ruled binder paper was as anachronistic a tool as a ruler for making a palm number line? Students could construct equally spaced parallel lines with a compass and a straightedge, but the construction would be tedious, and a sheet of lined paper can save a great deal of time.

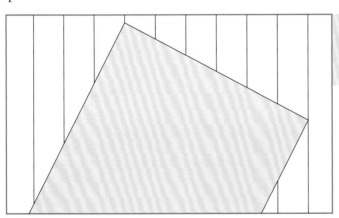

Fig. **1.2.**

Using lined binder paper and the 3-inch edge (≈ 1 palm) of a 3-by-5-inch index card to form six equal subdivisions of a palm

To find the length of $\sqrt{2}$ palms for the number line, a student might fold his or her index card as figure 1.3a shows. To find $\sqrt{3}$ palms, the student could use the length that he or she just obtained for $\sqrt{2}$ palms, along with the unit palm-length, as the legs of a right triangle, folding to obtain the hypotenuse of $\sqrt{3}$ palms, as figure 1.3b shows. Likewise, to determine the length of $\sqrt{5}$ palms, the student could form a right triangle with legs equal to $\sqrt{2}$ palms and $\sqrt{3}$ palms. Your students can generate many other square roots by the same method.

Fig. **1.3.**

Using 3-by-5-inch index cards to determine lengths for $\sqrt{2}$ palms and $\sqrt{3}$ palms

a. Using the 3-inch edge (≈ 1 palm) of a 3-by-5-inch index card to construct a fold-line of length $\sqrt{2}$ palms

b. Using the 3-inch edge (≈ 1 palm) of a second 3-by-5-inch index card, along with the length established in (a) for $\sqrt{2}$ palms, to construct a fold-line of length $\sqrt{3}$ palms

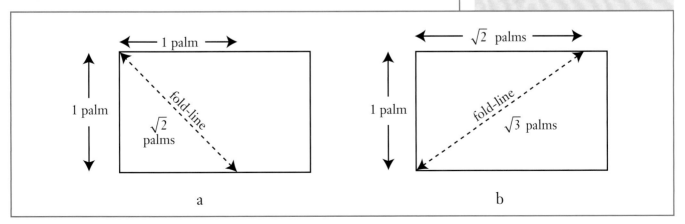

Chapter 1: The Real Numbers

← 1 palm →

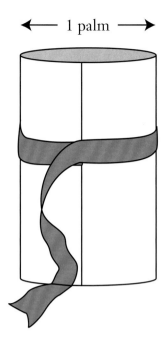

The students can readily find the length of π palms without a calculator or a ruler. They can roll a sheet of paper into a cylinder with a diameter of one palm. (Construction paper or tagboard, if either is available, will produce a cylinder of greater sturdiness.) By wrapping their number lines around the cylinder, starting at the 0 point and marking one complete revolution, students can find the length of π palms. See if your students can discover this method on their own.

Next, the activity introduces the students to operations on their palm-length number lines. In step 2, they investigate how to find lengths for $(a + b)$, $(a - b)$, $(a \times b)$, and $(a \div b)$ palms when they are given a and b. Specifically, they are asked to find lengths for $(\sqrt{2} + \pi)$ palms, $(\pi - \sqrt{2})$ palms, $(\pi \times \sqrt{2})$ palms, and $(\pi \div \sqrt{2})$ palms. To form the product $(\pi \times \sqrt{2})$ palms, students can construct a cylinder whose diameter is $\sqrt{2}$ palms, wrap their number lines one complete turn around it, starting at 0, and mark the resulting length in palms on their number lines. The new point will correspond to the cylinder's circumference, which is $(\pi \times \sqrt{2})$ palms.

Students may find the last calculation, $(\pi \div \sqrt{2})$ palms, to be more challenging than the others until they realize that it is equivalent to $(\pi \times \frac{\sqrt{2}}{2})$ palms:

$$\frac{\pi}{\sqrt{2}} \times \frac{\sqrt{2}}{\sqrt{2}} = \frac{\pi(\sqrt{2})}{2} = \pi \times \frac{\sqrt{2}}{2}.$$

Thus, they can accomplish the division by rolling a cylinder to a diameter of $\frac{\sqrt{2}}{2}$ palms and marking its circumference on their number lines.

Though the activity doesn't require a compass or a straightedge, steps 3–5 let the students encounter compass-and-straightedge methods for constructing lengths equal to $(a \times b)$ palms and $(a \div b)$ palms, given a and b. Working now in pairs, the students join their number lines at the 0 points to form a right angle. (They can use a corner of an index card to produce a right angle.) This arrangement allows them to simulate examples of compass-and-straightedge constructions of similar triangles (see the diagrams below). The students consider whether the triangles have to be right triangles, discovering that any pair of similar triangles will give the same results. They generalize their results with specific lengths to explain why the constructions yield $a \times b$ and $a \div b$ for any values.

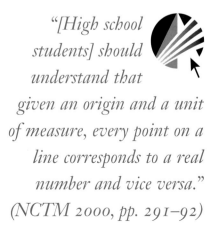

"[High school students] should understand that given an origin and a unit of measure, every point on a line corresponds to a real number and vice versa."
(NCTM 2000, pp. 291–92)

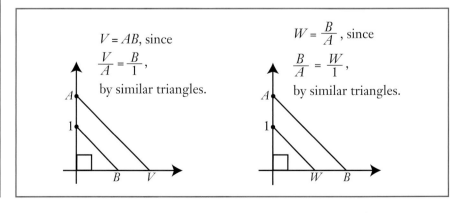

$V = AB$, since
$\frac{V}{A} = \frac{B}{1}$,
by similar triangles.

$W = \frac{B}{A}$, since
$\frac{B}{A} = \frac{W}{1}$,
by similar triangles.

You may decide that actually doing the constructions would be useful and appropriate for your students. If you wish, you can ask them to use a straightedge and a string compass to create similar triangles, as in the diagrams.

The students should grasp two key concepts:

1. All the real numbers—rational and irrational alike—can be thought of as representing points on a number line.
2. All the common arithmetic operations can be modeled geometrically.

This work with number lines can help the students develop a sense of number and operation that allows for treating rational and irrational numbers in the same way.

In step 6 of the activity, the students examine the properties of real number operations from a geometric point of view. As before, they use similar triangles, this time demonstrating that (a) the geometric method of multiplication is a commutative operation with 1 as the multiplicative identity and (b) any number except 0 has a multiplicative inverse under this operation.

The students are now set to think about how they might multiply two real numbers considered not as points with equal status on a line but as infinite, nonrepeating decimals. They consider numbers in this way in step 7, discussing how they could now perform operations and make calculations like those that they handled so neatly with geometry in step 2:

$$\frac{1+\sqrt{5}}{2}, \pi - \sqrt{2}, \pi \times \sqrt{2}, \text{ and } \frac{\pi}{\sqrt{2}}.$$

These discussions should help your students gain a new perspective on issues at the heart of real numbers and their operations.

Needless to say, no one expects the students to articulate a theory of infinite decimals in any formal way. However, as a result of their discussions, they should develop a better appreciation for the basic features of the real number system—the backbone of mathematics in grades 9–12.

Step 8 concludes the activity by returning briefly to the scenario with which the students began. They now consider what numbers, if any, they might want to add to their number lines to get them ready for the Ancient Egyptian Revival Society. For example, they might wish to mark out seven palms on their lines as the length that the Rhind Papyrus gives for one cubit.

The students might also research the history of irrational numbers, as well as the Pythagorean theorem, which they apply in creating their number line. Noting the eras of these discoveries can broaden the students' acquaintance with the history of mathematics while raising interesting questions of chronology in the context of the scenario. The Ancient Egyptian Revival Society wants the students' number lines to recreate a measuring system from 2000 B.C. Would the society question the authenticity of including irrational numbers, such as $\sqrt{2}$ and $\sqrt{3}$? Would the society regard these numbers as anachronisms on lines recreating the ancient Egyptian system of palms and cubits? Would the society consider using the Pythagorean theorem to be as anachronistic as using a ruler?

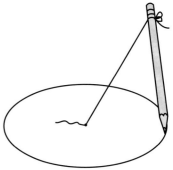

A string compass is a pencil tied to a long string. A student can draw a circle with a string compass by choosing a point on the string to be at the center, holding the string down at that point, and pulling the string taut while drawing around the point. This is a very ancient method of constructing a circle.

Historians often credit a Pythagorean, Hippasus of Metapontum (ca. 500 B.C.), with the discovery of irrational numbers. Earlier civilizations encountered some of these numbers without recognizing them as irrational. The ancient Egyptians, for example, knew that the ratio of a circle's circumference to its diameter was constant but supposed that it had a rational value, sometimes given as 256/81.

Legend has it that Hippasus's fellow Pythagoreans drowned him after he demonstrated that the square root of 2 could not be expressed as the ratio of two integers. Students can search the Web to find additional information on the history of irrational numbers at such sites as http://www.answers.com/topic/-irrational-number.

Though Pythagoras (ca. 569–500 B.C.) may have been the first to offer a proof of what we call the Pythagorean theorem, ancient tablets (ca. 1900–1600 B.C.) indicate that the Babylonians already knew of the relationship that the theorem identifies between the sides and the hypotenuse of a right triangle. The tablets also suggest that the Babylonians had rules for generating Pythagorean triples and approximated $\sqrt{2}$ accurately to five decimal places. Students can find information about the history of the Pythagorean theorem on the Web at such sites as http://www.ualr.edu/~lasmoller/pythag.html.

For an extension to Designing a Line, see the activity Pi Ruler in *Navigating through Measurement in Grades 6–8* (Bright et al. 2005, pp. 62–65; 128–30). In this activity, students create a measuring tool that that they use to make a direct measurement of a tree's circumference in centimeters and an indirect measurement of its diameter in π-scaled centimeters. They then use their results, along with established multipliers, to gauge the age of a tree.

Assessment

Be sure to monitor your students' efforts to devise geometric methods for constructing numbers on their number lines. Let the students share ideas and try a variety of strategies. Indeed, students can discover the methods that the text describes with just a few hints or leading questions. Whatever methods your students use, they should be able to justify them geometrically.

Several extensions can reveal how well students understand the essential ideas of the activity. You might ask your students to use the geometrical method introduced in step 3 to show that multiplication is associative and distributes over addition. You could also ask students to generalize the method to show how to multiply two numbers that might be negative.

You can also assess your students' understanding by asking them to apply what they have learned to a different problem in a new context. For example, how might someone determine the diameter of a tree quickly and effortlessly from a measurement of its circumference? Foresters use an instrument called a π-ruler for this task. On these rulers, the unit is π inches long. Thus, if the circumference of a tree measures six of these units, the forester concludes that the diameter of the tree is 6 inches. In general, circles whose circumferences measure n of these units have diameters of n inches.

To assess your students' understanding of the ideas developed in this activity, you could ask them to transform their number lines into π-rulers, with a unit equal to π palms, and with tenths of π palms marked across the number line, on its opposite edge. You could then have your students use their tapes to measure the diameters, in palms, of various cylindrical objects—even trees.

Where to Go Next in Instruction

The activity Designing a Line shows rational and irrational numbers in the same light—as points on a line. The next activity, Trigonometric Target Practice, provides a geometrical way to visualize the differences between rational and irrational numbers. It also connects the idea of a rational number to a lattice point in the Cartesian plane.

Trigonometric Target Practice

Goals

- Explore the values of the tangent function graphically to discover that tan θ is irrational for all rational values of θ, except θ = 0, when θ is in radians
- Discover that no line through the origin with an irrational slope will ever contain a lattice point other than the origin

Materials and Equipment

For each student—

- A copy of the activity sheet "Trigonometric Target Practice"
- A graphing calculator

pp. 89–90

Discussion

Students begin the activity by examining the graph of a line *l* that passes through the origin and the point (h, k), making an angle of θ radians with the positive *x*-axis. The graph is similar to that in fig. 1.4. Step 1 of the activity directs the students to consider the tangent of θ and write the equation of the line. These tasks are preliminaries to analyzing the relationship between the slope of a line through the origin and the angle that the line makes with the positive *x*-axis.

For the activity, the students should understand that in a right triangle the tangent of either acute angle is the ratio of the length of the opposite leg to the length of the adjacent leg. Thus, when they consider how to calculate the slope of line *l*, they will understand that the slope equals the tangent of the angle that the line makes with the positive *x*-axis. This awareness should enable them to discuss the connection between the angles for which the tangent function is undefined and the lines that have no slope. (See fig. 1.4.)

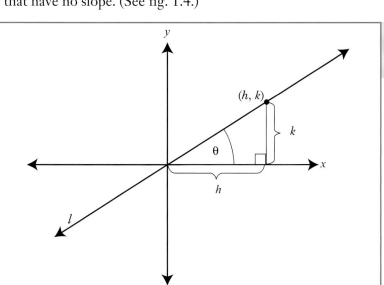

Fig. **1.4.**
The slope of the line, or $\frac{k}{h}$, is the tangent of θ

Chapter 1: The Real Numbers

A *lattice point* is a point that falls at the intersection of grid lines on a Cartesian grid. Thus, the coordinates of a lattice point are integers.

Step 2 of the activity introduces the idea of a *lattice point* and asks the students to determine that for a line to pass through the origin, (0, 0), and a second lattice point, (p, q), where $p \neq 0$, the slope of the line must be $\frac{q-0}{p-0}$, or $\frac{q}{p}$. Furthermore, they discover that the line also passes through all lattice points (r, s) such that $\frac{s}{r}$ is equivalent to $\frac{q}{p}$.

After establishing connections among the angle θ, the slope of the line, and tan θ, the students go on to explore graphs of $y = \tan(\theta) \cdot x$, with θ measured in radians. Step 3 instructs them to enter various rational fractions as values for θ and use the graphs that their calculators generate to investigate whether the resulting lines "hit" a lattice point other than (0, 0). This process of selecting and testing rational fractions will discourage students from using the calculator's arctangent function or its π key (or rational fractions that are equivalent to π up to the precision of the calculator).

Your students may eagerly compete with one another to find a value for θ that yields a line that "hits" a lattice point. In fact, no one can succeed without "tricking" the calculator, creating an appearance of success that will not hold up to scrutiny. After a short time, the students will either give up or claim they have found a value of θ and a lattice point (h, k) with the necessary property.

Any student who suggests a value for a possible "hit" should allow his or her classmates to put it to a stringent graphing-calculator test. As a class, the students should verify the proposed value and resulting line by using a sequence of "zoom" images to show that no matter how far in on the line the calculator zooms, the graph continues to represent the line as hitting the lattice point. Figure 1.5a shows a line that seems to be a candidate initially, apparently passing through the point (2, 3) as well as the origin. However, as figure 1.5b shows, zooming in on (2, 3) reveals space between the line and the point.

Fig. **1.5.**

"Zooming in" with a graphing calculator to test a graph of $y = \tan(1) \cdot x$ to see if the line passes through the point (2, 3)

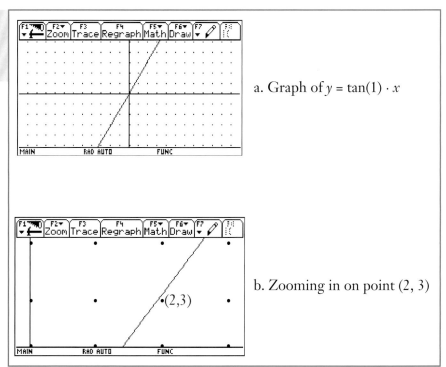

a. Graph of $y = \tan(1) \cdot x$

b. Zooming in on point (2, 3)

Tricking the calculator to appear to succeed in hitting lattice points is easy enough to do. Students can use several different methods. For example, the graph of $y = \tan\left(\frac{\pi}{4}\right) \cdot x$, or $y = x$, passes through all lattice points of the form (n, n) for integers n. Thus, if calculator users enter

$$y = \tan\left(\frac{3141592653589}{400000000000}\right) \cdot x,$$

which is same as the equation $y = \tan\left(\frac{\pi}{4}\right) \cdot x$ up to the precision of many calculators, then the graph of the line will appear to hit all the lattice points of the form (n, n) for integers n. Zooming in on these points will show no space between the line and the lattice point since the lattice point satisfies the equation of the line up to the precision of the calculator. Likewise, students might enter arctan $\frac{3}{4}$ for θ, or a rational fraction equivalent to the value of this arctangent, up to the precision of the calculator. This strategy will generate a line that goes through (4, 3) and many more lattice points on the calculator's screen. Zooming in on (4, 3) will not alter the appearance that the line passes through (4, 3).

In step 4, the final step of the activity, students generalize their results to conclude that when θ is a rational number of radians, tan θ is always irrational, except when θ = 0. In fact, if students complete the activity Solve That Number in chapter 2, they will appreciate the fact that the values of tan θ are not only irrational but also *transcendental* when θ is a rational number of radians, with the exception of θ = 0.

Assessment

What would happen in this activity if the students measured θ in degrees instead of radians? You can use this question as an extension that will allow you to assess how well your students have grasped the main ideas of the activity. In degrees, except when θ = 0° ± 180n, or when θ = 45° ± 90n, tan θ is irrational for integers n when θ is rational (Niven 1964).

Other extensions can work equally well as opportunities for assessment. In general, mathematics calls tan x, sin x, cos x, log x, ln x, and exp x *transcendental functions*, since for some rational numbers x, these functions have values that are transcendental. In fact, their output values are always transcendental for rational numbers x, except in the obvious cases, such as $x = 0$. You could have your students use any of these functions in place of tan θ to determine the slopes of lines through the origin. The results would be largely the same as for tan θ, with some exceptions as a result of the unique characteristics of each function.

For example, for fixed values of θ, the line $y = \sin(\theta) \cdot x$ has a slope of sin θ, but the angle formed by the line with the positive x-axis does not have measure θ as in the case of $y = \tan(\theta) \cdot x$. Using such functions not only lets you assess your students' understanding of the ideas in the activity but also helps students realize how prevalent irrational numbers are in the system of real numbers.

Be sure that your students avoid a common pitfall in the activity. After completing their work, some students might mistakenly suppose that no nonzero values of θ, rational or irrational, exist for which the

> The real numbers, which we customarily sort into rational and irrational numbers, are also divisible into *algebraic* and *transcendental* numbers. Every transcendental number is irrational, but some irrational numbers are algebraic. A number is algebraic if and only if it is the root of an algebraic equation
>
> $a_0x^n + a_1x^{n-1} + \cdots + a_{n-1}x + a_n = 0$
>
> where $a_0 \neq 0$, each a_i is an integer, and n is a natural number. Thus, the ratio π, for example, is transcendental. Chapter 2 explores the distinction between algebraic and transcendental numbers in the activity Solve That Number.

graph of $y = \tan(\theta) \cdot x$ passes through lattice points. Assess your students to be certain that they understand that to hit a lattice point (h, k), where h and k are nonzero, θ must be irrational. One value that works is $\theta = \arctan\left(\dfrac{k}{h}\right)$.

Conclusion

Understanding the real number system is one of the critical achievements of students in grades 9–12. Chapter 1 has focused on the real number system, especially highlighting issues related to the differences between the rational and irrational numbers.

Chapter 2 introduces activities that present a variety of issues in number theory and algebra. These issues include the infinitude of prime numbers, the theory of equations, definitions of algebraic and transcendental numbers, and interpretations of the complex numbers.

NAVIGATIONS SERIES

GRADES 9–12

Navigating through Number and Operations

Chapter 2
Number Theory and Algebra

"What can you say about the number that results when you subtract 1 from the square of an odd number?"
(NCTM 2000, p. 292)

Students in grades 9–12 are fully prepared to explore and prove some of the properties of the number and operation systems with which they have worked. Numerous properties of the integers, for instance, are now accessible to them. *Principles and Standards for School Mathematics* (NCTM 2000, p. 292) describes one possible investigation, in which students examine the numbers formed by subtracting 1 from the square of an odd number. After considering a few numbers, the students may observe that these numbers are divisible by 8. They can then use algebraic reasoning to show that such numbers are always divisible by 8. Likewise, students in grades 9–12 can investigate significant properties of real and complex numbers and their operations.

This chapter includes three activities that focus on systemic properties of integers, real numbers, and complex numbers. The first activity, Counting Primes, helps students reason that consecutive integers can have no common prime factors. The students then use this fact to prove that the set of prime numbers is infinite. In the second activity, Complex Numbers and Matrices, students explore a one-to-one correspondence between the set of complex numbers and a special set of 2×2 matrices. They prove that the set of matrices has essentially the same properties as the complex numbers under the operations of multiplication and addition. The third activity, Solve That Number, shows students the difference between algebraic and transcendental numbers. As they work, they also discover some very important properties of complex numbers and their conjugates.

"High school students should understand more fully the concept of a number system, how different number systems are related, and whether the properties of one system hold in another. Their increased ability to use algebraic symbolism will enable them to make generalizations about properties of numbers that they might discover. They can study and use vectors and matrices."
(NCTM 2000, p. 291)

The activities in this chapter are similar to many that teachers use throughout the 9–12 mathematics curriculum. However, the inclusion of these particular activities reflects the fact that they point toward large topics of major importance in mathematics. Counting Primes offers points of departure to many central ideas of number theory in the areas of divisibility, prime numbers, and prime factorization. Complex Numbers and Matrices gives students a look at the properties of an entire system of numbers and operations while illustrating the flexibility of matrix representations. Solve That Number introduces students to the important arena that mathematics has traditionally called the theory of equations.

All the activities emphasize the role of technology in the study of number and operation. Counting Primes illustrates how students can use technology appropriately in the prime factorizations of large numbers. Complex Numbers and Matrices relies on technology for computing the products and sums of matrices. Solve That Number shows students how to use technology to evaluate polynomials in x when substituting complex numbers for x.

A computer algebra system (CAS) would provide excellent enrichment and support in all cases. However, students can do each activity effectively with ordinary graphing calculators. In keeping with the spirit of *Principles and Standards*, the authors recommend that students do many of the computations by hand, using technology to search multiple examples for patterns and generalize results.

Counting Primes

Goals

- Learn how to determine whether a number is prime or composite
- Prove that two consecutive integers have no positive common factor except 1
- Explain why the set of primes is infinite

Materials and Equipment

For each student—
- A copy of the activity sheet "Counting Primes"
- A graphing calculator
- (Optional) Access to a computer algebra system (CAS)

pp. 91–94

Discussion

Two integers are *relatively prime* if their greatest common divisor is 1. In this activity, students prove that any two consecutive integers are relatively prime. They use this result to prove the following proposition: "When $p_1, p_2, ..., p_n$ are prime numbers, the prime factorization of $p_1 \times p_2 \times ... \times p_n + 1$ contains none of the primes $p_1, p_2, ..., p_n$." This insight prepares the students to discover that no finite set of prime numbers can contain all the prime numbers. Thus, they prove that the set of prime numbers is infinite by a method similar to the one that Euclid used in the fourth century B.C. (Beiler 1966, p. 212).

Some students may need a review of the definitions of *multiple*, *divisor*, *factor*, *prime*, and *relatively prime*. Afterward, as a warm-up activity, you can have each student in the class randomly select two positive integers and determine whether they are relatively prime. Pool the class results and determine the experimental probability that two randomly chosen integers are relatively prime.

The theoretical probability of this event is $\frac{6}{\pi^2}$ (Beiler 1966, p. 225). The proof of this result is related to perfect squares (Kalman 1993)—a topic probed in an activity in chapter 4—and rests on a formula discovered in the eighteenth century by Euler (Beckmann 1971, pp. 152–3):

$$\frac{6}{\pi^2} = \left(\frac{1}{1^2} + \frac{1}{2^2} + \frac{1}{3^2} + ... \right)^{-1}.$$

Students begin the activity Counting Primes by studying the spacing of the multiples of 5 on the number line. They use this spacing in step 1 to explain why two consecutive integers, which differ by 1, cannot both be multiples of 5. They then prove that the difference between two multiples of 5 is also a multiple of 5.

In step 2, they repeat the process with an integer other than 5. They compare their results with those of their classmates and generalize their findings to express the idea that the only common positive divisor of two consecutive integers is 1.

Step 3 asks students to consider two consecutive integers, M and $M + 1$, on the number line. The students must explain whether any positive

As a warm-up activity, you can have each student in the class randomly select two positive integers and determine whether they are relatively prime.

Chapter 2: Number Theory and Algebra

"All students should ... use number-theory arguments to justify relationships involving whole numbers." (NCTM 2000, p. 290)

integer(s) besides 1 can be a divisor of both numbers. It is important to let students discuss their justifications for saying that the greatest common divisor of two consecutive integers is 1.

Encourage alternative approaches. Some students might use geometric reasoning based on the number line, noting that multiples of a number n must be at least n units apart on the line, and hence the only value of n that would have multiples one unit apart is 1. Other students might attempt an algebraic argument: If M and N are both multiples of d, then $M = d \times h$ and $N = d \times k$, for some integers h and k. This result implies that $M - N = d \times h - d \times k = d(h - k)$. In other words, the difference between M and N is also a multiple of d. But if M and N are consecutive integers, then their difference is 1, and hence their only common positive divisor must also be 1.

In step 4, the students apply the definition of prime number by circling primes on a number line. Some students might circle negative numbers, thinking that they are also prime. Others might circle the number 1. Have the students refer to the definition on the activity sheet to resolve discrepancies in their lists. By definition, prime numbers must be greater than 1 and have exactly two positive divisors.

Before your students reach step 5, you should determine whether they need to review the idea of prime factorization or the fundamental theorem of arithmetic, which states that every natural number greater than 1 has a unique prime factorization. These concepts will be useful to them in the remainder of the activity.

Step 5 directs the students to select up to six numbers from the primes that they circled on the number line in step 4. They determine the product of these primes and add 1 to make a new number, N. They then use a calculator or other appropriate technology to find the prime factorization of N.

The students examine the characteristics of N in step 6. N might itself be a prime, though not one of the primes that the students selected. Or it might be a composite; the students might realize that if 2 is not one of their selected primes, then the N will be even and hence composite.

Since the students start with different sets of prime numbers, they are likely to make a variety of observations. However, they all should discover two important facts:

- No prime in their original list divides N.
- Whenever they divide N by any of their original primes, they obtain a remainder of 1.

The students can of course find the remainder though an integer division with a calculator. However, they will find their work in step 5 more enlightening if they divide by hand or argue the case by using a number line. Their process in one or two instances will make it very apparent that the remainder is always 1 when the divisor is any prime number from the original list of primes (see fig. 2.1).

Fig. **2.1.**
A by-hand computation showing that the remainder of $2 \times 3 \times 7 \times 13 \times 19 + 1$ divided by 13 is 1

$$\frac{2 \times 3 \times 7 \times 13 \times 19 + 1}{13} = \frac{2 \times 3 \times 7 \times 13 \times 19}{13} + \frac{1}{13} = 2 \times 3 \times 7 \times 19 + \frac{1}{13}$$

Steps 7 and 8 conclude the activity, helping the students realize that for any finite set of prime numbers, they can construct new prime numbers that do not belong to the set. This fact ensures that the set of prime numbers is not finite. In these steps, the students essentially follow the method of Euclid. Given any finite set of primes $S = \{p_1, p_2, p_3, ..., p_n\}$ they know $p_1 \times p_2 \times p_3 \times ... \times p_n$ and $p_1 \times p_2 \times p_3 \times ... \times p_n + 1$ are consecutive integers and therefore, by the results of step 3, have no common factors greater than 1. But the prime factors of $p_1 \times p_2 \times p_3 \times ... \times p_n$ are exactly the elements of S. Therefore, the set of prime factors of $p_1 \times p_2 \times p_3 \times ... \times p_n + 1$, which is nonempty, has no elements in common with S. Thus, S does not contain all the prime numbers.

Assessment

The activity Counting Primes pivots on the fact that two consecutive integers are relatively prime—that is, their greatest common divisor is 1. Specifically, the students discover that if $p_1, p_2, p_3, ..., p_n$ are prime numbers, then $p_1 \times p_2 \times p_3 \times ... \times p_n$ and $p_1 \times p_2 \times p_3 \times ... \times p_n + 1$ have no common prime factors. To assess your students' understanding of this idea, you might ask them to form other numbers that are relatively prime to $p_1 \times p_2 \times ... \times p_n$. Students should realize that the integer $p_1 \times p_2 \times p_3 \times ... \times p_n - 1$ is also relatively prime to $p_1 \times p_2 \times ... \times p_n$. But there are many other possibilities that the students can construct as well. For example, if m is any positive integer that is less than the smallest prime in the set $S = \{p_1, p_2, p_3, ..., p_n\}$, then $p_1 \times p_2 \times ... \times p_n + m$ is relatively prime to $p_1 \times p_2 \times ... \times p_n$. Or if q is a prime number not contained in S, then $p_1 \times p_2 \times ... \times p_n + q$ is relatively prime to $p_1 \times p_2 \times ... \times p_n$.

Some students may not understand the argument that the set of primes is infinite. Ask them to investigate the claim that if they began with 2 and made an ordered list of all the prime numbers, no matter how long their list became, there would always be prime numbers larger than any prime on their list. They can work with a calculator (TI-92 or Voyage 200, for example), a Casio Classpad 300, or an online factorization program (see fig. 2.2) to extend a list such as that in table 2.1 to include many more of the initial primes.

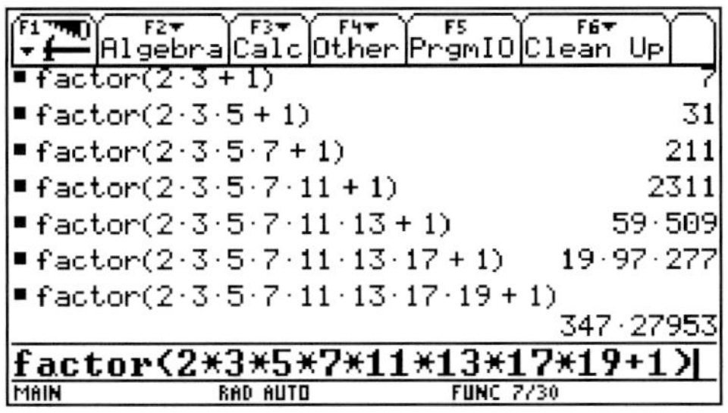

Fig. **2.2.**

Using technology to extend the pattern to longer lists of primes

Table 2.1
Representations of $N = p_1 \times p_2 \times ... \times p_n + 1$, for Increasing Ordered Lists of Primes

Prime Numbers	Expression for N
2, 3	$2 \times 3 + 1 =$
2, 3, 5	$2 \times 3 \times 5 + 1 =$
2, 3, 5, 7	$2 \times 3 \times 5 \times 7 + 1 =$

Chapter 2: Number Theory and Algebra

Note that if students go so far as to list all the primes from 2 to 67, the process will become awkward, since the smallest factor of $2 \times 3 \times 5 \times 7 \times 11 \times \ldots \times 59 \times 61 \times 67 + 1$ is 54,730,729,297. The factoring algorithms on even the most powerful calculators must search through too many prime numbers to find this prime factor in a reasonable time.

It will come as no surprise to many students that there are infinitely many primes. After all, they know that there are infinitely many integers, so they can easily imagine that there are infinitely many primes. Nevertheless, to appreciate the significance of—and the need for proving—the infinitude of primes, students should understand that prime numbers become increasingly sparse farther out on the number line. This fact implies that arbitrarily large gaps exist between consecutive primes on the number line.

You can assign a variety of tasks to assess your students' understanding of the main ideas in the activity. You could ask the students to generate, for any positive integer n, a string of $n - 1$ consecutive numbers, all of which are composite. They could prove that all the numbers in the set $\{n! + 2, n! + 3, \ldots, n! + n\}$ are composite. For example, if you gave them $n = 101$, they would create 100 consecutive composite numbers.

Alternatively, you could ask your students to show that for

$$2 \times 3 \times 5 \times 7 \times \ldots \times p_n + m,$$

where p_n is the nth prime number and $1 < m < p_{n+1}$, they can generate a string of $(p_{n+1} - 2)$ consecutive numbers, all of which are composite. For example, consider the product of the first eleven prime numbers: $a = 2 \times 3 \times 5 \times 7 \times 11 \times \ldots \times 31$. The twelfth prime number is 37. Note that $a + 2, a + 3, a + 4, \ldots,$ and $a + 36$ form a string of 35 consecutive composite numbers. There is no doubt that each of these numbers is composite, since the numbers from 2 to 36 all have prime factors that are less than 37, and all such prime factors are also factors of a.

Where to Go Next in Instruction

Your students might be interested to know that though prime numbers become increasingly rare as the number line continues, a prime always exists between x and $2x$ for every integer x greater than 3. In 1850, Chebyshev proved this fact, known as Bertrand's conjecture (Beiler 1966, p. 227; Francis 1993, p. 88).

Your students can find many Web sites that are devoted to prime numbers. These include interesting sites related to the search for larger and larger prime numbers. This endeavor is of interest in such applications as cryptography, where prime numbers have special usefulness, as chapter 3 demonstrates in the activity Rock Around the Clock.

Counting Primes gives students an opportunity to reason about key ideas in number and operations. Thinking about multiples, divisors, prime factors, relatively prime numbers, and the infinitude of primes can deepen students' understanding of the integers while providing excellent settings for reasoning and proof. The next activity, Complex Numbers and Matrices, allows students to extend their reasoning by establishing field properties for a special set of matrices.

Useful information about prime numbers appears at a variety of Web sites, including the following:

- http://www.utm.edu/research/primes/

(includes links to numerous Web sites with up-to-date information about prime numbers and the results of mathematics research focusing on them)

- http://mathforum.org/dr.math/faq/faq.prime.num.html

("Ask Dr. Math" answers questions and provides many links to useful resources for students and mathematics classrooms)

- http://www-groups.dcs.st-and.ac.uk/~history/HistTopics/Prime_numbers.html

(provides a short history of prime numbers)

Complex Numbers and Matrices

Goals

- Demonstrate fluency with addition and multiplication of complex numbers
- Understand different representations of complex numbers, including ordered pairs, vectors, and matrices
- Determine properties of the complex number system and compare them with the properties of other systems

Materials and Equipment

For each student—
- A copy of the activity sheet "Adding Complex Numbers"
- A copy of the activity sheet "Multiplying Complex Numbers"
- A calculator that can multiply matrices
- (Optional) Access to a computer algebra system (CAS)

pp. 95–97; 98–100

Discussion

In this activity, students connect four related representations of complex numbers. They explore the usual representation of complex numbers as numbers of the form $a + bi$, as well as the common geometric representation of them as points in the complex plane (ordered pairs of real numbers) and as vectors. They then encounter an unusual fourth representation of the complex numbers—as matrices of the form $\begin{bmatrix} a & b \\ -b & a \end{bmatrix}$, where a and b are real numbers.

The activity guides the students in exploring the operations of addition and multiplication, looking for patterns, and making connections among the various representations. Students should already be acquainted with addition and multiplication of complex numbers and matrices before beginning the activity.

Representing complex numbers with matrices may seem strange to your students. However, as a system of numbers and operations, the matrix representation satisfies all the properties of the field of complex numbers. Figure 2.3 shows that the set of matrices of the form $\begin{bmatrix} a & b \\ -b & a \end{bmatrix}$, where a and b are real numbers, is closed under addition and multiplication. Thus, if two matrices of this form are added or multiplied, the result is a matrix of the same form.

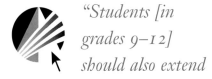

"Students [in grades 9–12] should also extend their understanding of operations to number systems that are new to them." (NCTM 2000, p. 293)

A set is *closed* under an operation if performing the operation on two elements of the set produces another element of the set.

$$\begin{bmatrix} a & b \\ -b & a \end{bmatrix} + \begin{bmatrix} c & d \\ -d & c \end{bmatrix} = \begin{bmatrix} a+c & b+d \\ -b-d & a+c \end{bmatrix}$$

$$\begin{bmatrix} a & b \\ -b & a \end{bmatrix} \cdot \begin{bmatrix} c & d \\ -d & c \end{bmatrix} = \begin{bmatrix} a \cdot c - b \cdot d & a \cdot d + b \cdot c \\ -a \cdot d - b \cdot c & a \cdot c - b \cdot d \end{bmatrix}$$

Fig. **2.3.**

Closure of the set of matrices of the form $\begin{bmatrix} a & b \\ -b & a \end{bmatrix}$ under addition and multiplication

Chapter 2: Number Theory and Algebra

A one-to-one correspondence exists between matrices of the form $\begin{bmatrix} a & b \\ -b & a \end{bmatrix}$ and the complex numbers. Figure 2.4 demonstrates this correspondence, under which the operations of multiplication and addition are preserved. When such a correspondence exists between two algebraic systems of number and operations, the two systems are isomorphic. Since $\begin{bmatrix} 1 & 0 \\ 0 & 1 \end{bmatrix}$ is the multiplicative identity of the 2 × 2 matrices, it is natural for this matrix to correspond to 1, or $1 + 0i$, which is the multiplicative identity of the complex numbers. Likewise, since $\begin{bmatrix} 0 & 1 \\ -1 & 0 \end{bmatrix}^2 = \begin{bmatrix} -1 & 0 \\ 0 & -1 \end{bmatrix}$, the additive inverse of $\begin{bmatrix} 1 & 0 \\ 0 & 1 \end{bmatrix}$, it is natural for the matrix $\begin{bmatrix} 0 & 1 \\ -1 & 0 \end{bmatrix}$ to correspond to the complex number i, which has the property $i^2 = -1$, which is the additive inverse of 1.

Fig. **2.4.**
A one-to-one correspondence between matrices of the form $\begin{bmatrix} a & b \\ -b & a \end{bmatrix}$ and complex numbers of the form $a + bi$

$$\begin{bmatrix} 1 & 0 \\ 0 & 1 \end{bmatrix} \xleftrightarrow{\text{corresponds to}} 1 \qquad \begin{bmatrix} 0 & 1 \\ -1 & 0 \end{bmatrix} \xleftrightarrow{\text{corresponds to}} i$$

$$\begin{bmatrix} a & b \\ -b & a \end{bmatrix} = a\begin{bmatrix} 1 & 0 \\ 0 & 1 \end{bmatrix} + b\begin{bmatrix} 0 & 1 \\ -1 & 0 \end{bmatrix} \xleftrightarrow{\text{corresponds to}} a(1) + b(i)$$

Historically, complex numbers provided a means for solving quadratic and other polynomial equations. Although "imaginary" numbers appeared in algebra in the sixteenth century, mathematicians largely discounted them until the early nineteenth century, when Wessel and Argand introduced their geometric representation as points in the complex plane (Boyer 1968, p. 548).

Part 1, "Adding Complex Numbers," introduces students to this representation and explores adding complex numbers—first as vectors and then as 2 × 2 matrices. Part 2, "Multiplying Complex Numbers," pursues the use of 2 × 2 matrices to represent complex numbers, now focusing on their multiplication.

If your students have not encountered vectors, or if they need to review them, you can have them work with the vector activities on the CD-ROM, "Driving with One Vector" and "Flying with Two Vectors." Both of these activities include applets that demonstrate the influence of the magnitude and the direction of a vector.

You can use the applet activities "Driving with One Vector" and "Flying with Two Vectors" on the CD-ROM to introduce your students to vectors or extend their understanding of them.

Part 1—"Adding Complex Numbers"

Part 1 of Complex Numbers and Matrices introduces students to the geometric representations of complex numbers as vectors and as points in the plane denoted by ordered pairs of real numbers. After a quick review of adding complex numbers in step 1, students investigate addition with both ordered pairs and vectors in steps 2 and 3 to establish the connection in each case with the addition of complex numbers. To avoid confusion in notation, this brief activity does not use the vector notation $a \cdot \hat{\imath} + b \cdot \hat{\jmath}$ but instead has students construct parallelograms whose diagonal vector is the sum of two vectors.

In step 4, students examine addition of matrices of the form $\begin{bmatrix} a & b \\ -b & a \end{bmatrix}$.

Step 5 concludes the activity by having the students show that the set of matrices of this form is closed under addition. In this step, the students also establish the correspondence between the addition of complex numbers and the addition of matrices of this form.

Part 2—"Multiplying Complex Numbers"

In part 2 of Complex Numbers and Matrices, students delve deeper into the relationship between complex numbers ($a + bi$) and matrices of the form $\begin{bmatrix} a & b \\ -b & a \end{bmatrix}$ by investigating multiplication. The students find the products of three pairs of complex numbers in step 1, and in step 2 they multiply three pairs of matrices. After your students have completed one computation of each type by hand, you might decide to have them use a calculator for others of the same type so that they can begin to investigate the details and establish patterns.

Your students will continue this work in step 3, which asks them to use examples of their own to compare their work in steps 1 and 2. After constructing their own examples, they can quickly check the results. Whether they work with a calculator, with a CAS (as illustrated in fig. 2.5), or by hand, they should be able to generalize their results to establish the one-to-one correspondence between the product of two matrices of the form $\begin{bmatrix} a & b \\ -b & a \end{bmatrix}$ and the product of two complex numbers.

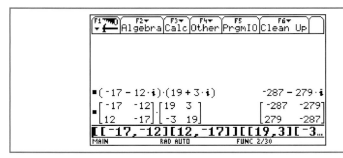

Fig. **2.5.**

Using a CAS to discover and validate the one-to-one correspondence between the product of two matrices of the form $\begin{bmatrix} a & b \\ -b & a \end{bmatrix}$ and the product of two complex numbers

In general, matrix multiplication is not commutative. Step 4 reminds the students of this fact by having them consider the multiplication of two 2 × 2 matrices that are not of the form $\begin{bmatrix} a & b \\ -b & a \end{bmatrix}$. However, step 5 leads them to conclude that multiplication for matrices of the form $\begin{bmatrix} a & b \\ -b & a \end{bmatrix}$ is commutative. Figure 2.6 illustrates how technology can

Fig. **2.6.**

Using a CAS to investigate and validate the commutative property for the multiplication of 2 × 2 matrices of the form $\begin{bmatrix} a & b \\ -b & a \end{bmatrix}$

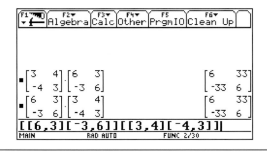

"All students [in grades 9–12] should ... develop fluency in operations with real numbers, vectors, and matrices, using mental computation or paper-and-pencil calculations for simple cases and technology for more complicated cases."
(NCTM 2000, p. 290)

facilitate this discovery by exploring many specific cases as well as the general case. You should also encourage your students to attempt paper-and-pencil demonstrations of the results since multiplication of 2×2 matrices is relatively simple.

In step 6, the students compute the multiplicative inverses for a complex number and a matrix of the form $\begin{bmatrix} a & b \\ -b & a \end{bmatrix}$. Once again, they are encouraged to notice the correspondence, this time between the inverses. To find the inverses, the students multiply the given expressions and set up a system of equations. The students must solve for a and b in the equation $(3 + 4i)(a + bi) = (1 + 0i)$. Multiplying yields $(3a - 4b) + (4a + 3b)i = (1 + 0i)$, and this equation yields the system

$$\begin{array}{l} 3a - 4b = 1 \\ 4a + 3b = 0 \end{array}.$$

Likewise, $\begin{bmatrix} 3 & 4 \\ -4 & 3 \end{bmatrix} \begin{bmatrix} a & b \\ -b & a \end{bmatrix} = \begin{bmatrix} 1 & 0 \\ 0 & 1 \end{bmatrix}$ yields $\begin{bmatrix} 3a - 4b & 4a + 3b \\ -(4a + 3b) & 3a - 4b \end{bmatrix} = \begin{bmatrix} 1 & 0 \\ 0 & 1 \end{bmatrix}$,

which implies the same system. Solving the system of equations, students get $a = \dfrac{3}{25}$ and $b = \dfrac{-4}{25}$. They can confirm this on a calculator and generalize their results as in figure 2.7.

Fig. 2.7.
Using a CAS to investigate and generalize multiplicative inverses

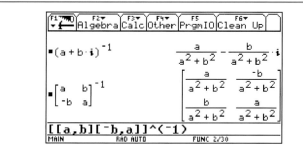

Step 7 concludes part 2 of the activity with an investigation of the correspondence between the properties of the complex numbers and matrices of the form $\begin{bmatrix} a & b \\ -b & a \end{bmatrix}$ under addition and multiplication. This step gives students a chance to review the field properties of these operations. For example, students should check to see whether multiplication is associative, whether multiplicative and additive inverses exist for all nonzero $\begin{bmatrix} a & b \\ -b & a \end{bmatrix}$, and whether multiplication distributes over addition.

Assessment

The final step of part 2 offers an excellent opportunity to assess students' understanding of the correspondence between the complex numbers and the set of matrices of the form $\begin{bmatrix} a & b \\ -b & a \end{bmatrix}$. You can divide your students into groups and ask the groups to present their results orally to the class. Having the students make presentations allows for an easy and efficient check of their understanding of the properties of these systems of numbers and operations.

You may find that some students are not ready to compare the systems property by property because they cannot articulate what the significant properties are. However, on the basis of their work in the activity, even these students should be able to say that the systems are essentially the same. They should observe that the elements in one match the elements in the other, noting as well that when someone adds or multiplies these elements, the results match. This matching—or correspondence—is essentially the notion of isomorphism, which is so fundamental in mathematics.

There are many number systems that students in grades 9–12 can explore in the way that they investigated the matrix system in this activity. Such investigations can serve as very fruitful extensions of the activity, offering opportunities to assess students' understandings and misconceptions about the basic ideas of number systems.

Consider, for example, the following operations on the real numbers: $a \oplus b = a + b + 1$, and $a \otimes b = ab + a + b$. These operations give a system of numbers that is isomorphic to the real numbers except that the additive identity of the new system is -1, and its multiplicative identity is 0.

If you wish to give your students a more exotic example, you can challenge them to explore the following operation on the integers:

$$a \oplus b = \begin{cases} a+b & \text{if } a \text{ is even} \\ a-b & \text{if } a \text{ is odd} \end{cases}$$

They will discover that this operation is associative, with an identity element and inverses, but it is not commutative. When the operation is restricted to integers modulo n and n is even, it still has these properties, and the finite systems that it generates are isomorphic to the groups of symmetries of regular polygons with $\frac{n}{2}$ sides. When restricted to integers modulo n, where n is odd, the operation has vastly different properties and fails to be a group operation except in the trivial case where $n = 1$.

Where to Go Next in Instruction

The activity Complex Numbers and Matrices helps students make a profound connection between the field of complex numbers and the field of matrices of the form $\begin{bmatrix} a & b \\ -b & a \end{bmatrix}$. Along the way, they can improve their computational fluency with complex numbers and matrices. The next activity, Solve that Number, helps students deepen their understanding of complex numbers and real numbers by investigating numbers that are *algebraic*—that is, numbers that are the roots of polynomials with integer coefficients.

Solve That Number

Goals

- Deepen understanding of roots of polynomials
- Distinguish between algebraic and transcendental numbers
- Establish a connection between complex numbers and their conjugates
- Develop the connection between factors and roots of polynomials

Materials and Equipment

For each student—

- A copy of the activity sheet "Solving Real Numbers"
- A copy of the activity sheet "Solving Complex Numbers"
- A graphing calculator
- (Optional) Access to a computer algebra system (CAS)

pp. 101–2; 103–5

Discussion

The distinction between *algebraic* and *transcendental* numbers is important in mathematics. For example, the sine, cosine, and tangent functions are often called transcendental functions since $\sin x$, $\cos x$, and $\tan x$ are all transcendental numbers when x is any nonzero rational number of radians. This fact played a role in the activity Trigonometric Target Practice in chapter 1.

A number, whether real or complex, is algebraic if it is the root of a polynomial with integer coefficients. All other numbers are transcendental. The algebraic numbers form a field. More to the point, if you add, subtract, multiply, or divide (except by 0) two algebraic numbers, the result is also algebraic.

The activity Solve That Number introduces students to the idea of algebraic and transcendental numbers while helping them discover some important facts about roots of polynomials from the theory of equations. The activity is in two parts—"Solving Real Numbers" and "Solving Complex Numbers." In both parts, your students should be able to do the computations in several steps by hand. For example, they should use by-hand computations to evaluate quadratic polynomials, including cases in part 2, where x is replaced by a complex number.

Part 1—"Solving Real Numbers"

The first part of the activity explains that though students have often solved equations by finding the roots of polynomials, this time they will reverse the process to "solve" a root by finding a polynomial. They will start with a number x and search for an equation $P(x) = 0$, where $P(x)$ is a polynomial with integer coefficients and x is a root. In other words, x is a number that makes the polynomial equal to 0. The activity calls this process "solving the number."

Part 1 does not give the students a method for "solving" a real number. They must discover techniques on their own as they solve five numbers in step 1. For the number $1 + \sqrt{6}$, for example, the students must find

Adding, subtracting, multiplying, or dividing (except by 0) two algebraic numbers results in a number that is also algebraic.

an equation $P(1+\sqrt{6}) = 0$, where $P(x)$ is a polynomial with integer coefficients and with $1 + \sqrt{6}$ as a root. Some students will discover the polynomial $x^2 - 2x - 5$ by working backward from the quadratic formula. However, a simpler method is to let $x = 1 + \sqrt{6}$. Therefore, $x - 1 = \sqrt{6}$, and $(x - 1)^2 = 6$. This implies that $x^2 - 2x - 5 = 0$.

In step 2, the students compare the methods that they used to "solve" the numbers in step 1, and they consider whether one method works for all the numbers. The method outlined here for solving the number $1 + \sqrt{6}$ works quite well for the real numbers in part 1, and part 2 suggests this method explicitly for solving complex numbers.

Steps 3–5 of the activity invite the students to consider how to generate first just one additional "solution," and then many others, for each of the numbers in step 1. Steps 6 and 7 help the students connect all their work with the idea that a number is *algebraic* if it can be the root of a polynomial $P(x)$ with integer coefficients. The students demonstrate that when p and q are integers ($q \neq 0$), every rational number $\frac{p}{q}$ and every number of the form $p + \sqrt{q}$ is algebraic.

Students need to use algebraic reasoning to argue that any number of the form $p + \sqrt{q}$ is algebraic. If they approach the problem by setting $x = p + \sqrt{q}$, then they will get $(x - p)^2 = q$, or $x^2 - 2px + p^2 - q = 0$. Since p and q are integers, then so are $2p$ and $p^2 - q$. Thus, $x^2 - 2px + p^2 - q$ is a quadratic polynomial with integer coefficients that has $p + \sqrt{q}$ as a root. Note that its other root is $p - \sqrt{q}$.

All this work prepares the students to consider such numbers as π, which they cannot "solve." They realize, in other words, that there are numbers that cannot be roots of any polynomial $P(x)$ with integer coefficients. The students learn that such numbers are called *transcendental*. It is far beyond the scope of high school mathematics to prove that transcendental numbers cannot be solved. Indeed, proofs that some numbers are not the roots of any polynomial with integer coefficients only began to appear in the nineteenth century (Boyer 1968, p. 602).

Thus, step 8 asks the students only to list numbers that they think might be transcendental and then compare their lists. This exercise can promote lively discussions and interesting excursions on the World Wide Web, as well as explorations of efforts to use transcendental numbers to find other such numbers. The fact that the sum of a transcendental number and an algebraic number is transcendental is useful in this work.

Part 2—"Solving Complex Numbers"

In part 2, students extend the idea of "solving" a number to complex numbers. Before introducing this part of this activity, make sure that your students are ready for its main ideas. Have they had experience solving linear and quadratic equations? Do they know what it means for two complex numbers to be conjugates of each other? Some knowledge of the quadratic formula will also be useful, although it is not essential.

In step 1, the students "solve" pairs of numbers that are conjugates. Students who understand the quadratic formula will readily understand that if a complex number is a root of a quadratic polynomial with real coefficients, then so is its conjugate. This fact happens to be true for all polynomials with real coefficients, not just quadratics.

Interesting information about transcendental numbers appears at numerous Web sites, including the following:

- http://sprott.physics.wisc.edu/pickover/trans.html ("The 15 Most Famous Transcendental Numbers" by Cliff Pickover)

- http://mathworld.wolfram.com/TranscendentalNumber.html (from the developers of *Mathematica*, a list of transcendental numbers and when mathematicians proved them to be transcendental)

- http://mathforum.org/library/drmath/sets/high_transcendental.html ("Ask Dr. Math" answers questions and offers many links for useful resources for students and mathematics classrooms)

- http://www.cut-the-knot.org/do_you_know/numbers.shtml (information on different types of numbers, including algebraic and transcendental numbers, and links to related topics, in a general discussion of "What is a number?")

The *conjugate* of a complex number $a + bi$ is obtained by changing the sign of the imaginary part of the number. Thus, $a + bi$ and $a - bi$ are conjugates, and their product is $a^2 + b^2$.

In step 2, the students conjecture that this is true. They prove in step 3 that any complex number $p + qi$ is algebraic. In step 4, they work with and without technology to investigate the general claim that if $P(x)$ is a polynomial with integer coefficients, then $P(h + ki)$ and $P(h - ki)$ are conjugates. They extend this work in step 5, showing that if $P(x)$ is a polynomial with integer coefficients, then $P(a + bi) = 0$ if and only if $P(a - bi) = 0$.

By examining many cases with the help of technology, students can discover these facts on their own. Steps 4 and 5 encourage them to work with specific polynomials to demonstrate these ideas. Although a general proof for all polynomials would require the binomial theorem, students who have access to a computer algebra system can prove the cases for all quadratic and all cubic polynomials with relative ease. Figure 2.8 illustrates this process.

Fig. **2.8.**

Using a CAS to illustrate that the conjugate of $f(a + bi)$ is $f(a - bi)$ for polynomials with real coefficients

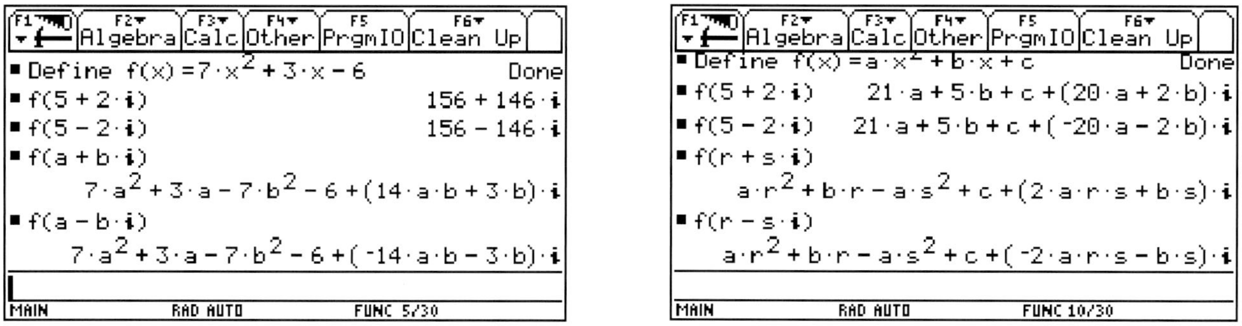

Assessment

It is important that you assess your students' arguments—sometimes while helping them reason about the problems at hand. For example, some students might have trouble "solving" a number like $1 + \sqrt{12}$. Depending on their knowledge of the quadratic formula, you might suggest that they assume that $1 + \sqrt{12}$ is a root of a quadratic polynomial of the form $x^2 + bx + c$ and try to determine values for b and c by substituting $1 + \sqrt{12}$ for x. Using this method, the students will discover that $(13 + b + c) + (2 + b)\sqrt{12} = 0$. By finding integer values for b and c so that $2 + b = 0$ and $13 + b + c = 0$ ($b = -2$, and $c = -11$), they can determine the requisite polynomial. The assumption that $1 + \sqrt{12}$ is a root of a quadratic polynomial of the form $x^2 + bx + c$ with integer coefficients thus leads to a solution. It turns out that this assumption is a valid conjecture for any number of the form $p + \sqrt{q}$ when p and q are integers, as the preceding discussion demonstrates.

Conclusion

Chapter 2 has focused on issues that arise within number theory and algebra. All the activities involve the students in proving general results. *Principles and Standards* emphasizes the importance of such theoretical investigations but also urges teachers to supplement them with activities that help students recognize the power and prevalence of applications in the world around them. Chapter 3 presents two interesting applications of number and operations.

NAVIGATIONS SERIES

GRADES 9–12

NAVIGATING through NUMBER and OPERATIONS

Chapter 3
Numbers and Operations in the World

Students in grades 9–12 should have opportunities to explore numbers and operations that are embedded in processes and instruments in the real world.

Students in grades 9–12 should have opportunities to explore numbers and operations that are embedded in the design of objects around them. According to the Curriculum Principle articulated in *Principles and Standards for School Mathematics* (NCTM 2000), teachers should create experiences that "allow students to see that mathematics has powerful uses in modeling and predicting real-world phenomena" (p. 16). Such experiences will be especially important as high school students encounter numbers and number systems that go well beyond those that they studied in the elementary grades. This chapter's two activities serve as samples of the kinds of investigations that you and your students can undertake of applications of number and operations in the physical world.

The first activity—Frequency, Scales, and Guitars—shows students that irrational numbers are useful in the design of musical scales and the construction of stringed instruments. Many teenagers enjoy music—particularly guitar music—so the mathematics embedded in modern stringed instruments may well interest and surprise your students. The ancient Greek understanding of stringed instruments fostered a theory of musical scales that used only rational numbers. In contrast, "well-tempered" scales, attributed to Johann Sebastian Bach and others, use $\sqrt[12]{2}$ to determine the frequencies of notes. Because the notes in well-tempered scales are equally spaced, musicians can transpose compositions easily to other keys. Yet, it is interesting that the frequencies of these well-tempered scales are hardly distinguishable from those of the Pythagorean scale.

In grades 9–12 students should be confidently using mathematics to explain complex applications in the outside world." (NCTM 2000, pp. 65–66)

The second activity, Rock Around the Clock, provides an excursion into the realm of coding and decoding secret messages. The mathematics in this activity, which involves modular arithmetic and the construction of finite number systems, is very accessible to students and will help them develop an understanding of other number systems, their operations, and their uses in coding and decoding communications.

Both activities rely on technology for data gathering, computation, and graphing. In fact, the applications of number and operation that the activities highlight are themselves part of the technology of the students' world. The activities lead the students through the nuts and bolts of the mathematical applications while emphasizing reasoning, multiple representations, connections, and communication.

It is important to assess students' understanding of the mathematics behind such applications. You will find suggestions for creative, open-ended extensions in the assessment sections in the text. These further explorations can illuminate your students' understanding of the main ideas of the activities themselves.

Frequency, Scales, and Guitars

Goals

- Compute the *period* and the *frequency* of sound waves
- Determine a model for constructing a musical scale
- Develop a model for building a stringed instrument

Materials and Equipment

For each student—

- A copy of the activity sheet "Frequency, Scales, and Guitars"
- A graphing calculator
- (Optional) Access to a guitar and a data-collection device (such as a Texas Instruments calculator-based laboratory [TI CBL 2] or a Vernier LabPro) with a microphone to measure sound waves

pp. 106–10

Discussion

Two musical notes are *consonant* if they make a pleasing sound when someone plays them together (Bibby 2003, p. 13). In ancient Greece, the Pythagoreans (ca. 550 B.C.) knew that the note that musicians produce by pressing a string down at its midpoint is especially consonant with the note that they produce on the string when it is "open"—when they do not press the string. In fact, the Pythagoreans and others before and after them designated an octave as the interval of sound between two such notes. Ancient musicians considered two notes that were one octave apart as equivalent in some sense, and they gave them the same name. They said that these notes differed by their pitch, with the higher note having twice the pitch of the lower note.

The Pythagoreans knew that when musicians pressed a string and produced a note on two-thirds of its length, the resulting note was also very consonant with the note that they produced on the open string. They designated the interval of sound between this note and the note on the open string as a perfect fifth, because it was the fifth note above the original note. They measured the pitch of this note as $\frac{3}{2}$ times the pitch of the open note. By starting at a note with pitch P and constructing five consecutive perfect fifths above P and one perfect fifth below P, we get a set of notes with the following pitches:

$$\left\{ P,\ \frac{3}{2} \times P,\ \frac{3}{2} \times \left(\frac{3}{2} \times P\right),\ \left(\frac{3}{2}\right)^3 \times P,\ \left(\frac{3}{2}\right)^4 \times P,\ \left(\frac{3}{2}\right)^5 \times P,\ \frac{2}{3} \times P \right\}$$

or

$$\left\{ P,\ \frac{3}{2}P,\ \frac{9}{4}P,\ \frac{27}{8}P,\ \frac{81}{16}P,\ \frac{243}{32}P,\ \frac{2}{3}P, \right\}.$$

Each musical note is a sound wave with a unique frequency. The frequency is the number of pulsations, or wave cycles, that the note produces in one second. Scientists measure frequency in hertz (Hz), or cycles per second. The applet activity Sound Wave on the CD-ROM gives a visual representation of sound waves in air.

Before the seventeenth century, scientists thought that two notes with different frequencies had different pitches. However, once scientists began to analyze sound as a wave, they began to identify pitch with frequency. Thus, from the standpoint of modern science, the last note in an octave has a frequency that is twice that of the first note in the octave.

Chapter 3: Number and Operations in the World

Centuries before the Pythagoreans, the Babylonians had divided the octave into scales of five tones (the *pentatonic scale*) and seven tones (the *septatonic scale*). The Pythagorean innovation was the attempt to explain the phenomenon of *consonance*, or harmony, by means of number and operation.

The notes corresponding to the last five pitches in this set do not belong to the octave between *P* and *2P*. However, by multiplying or dividing the other pitches by powers of 2, we can produce equivalent notes—equivalent in that they are a whole number of octaves apart—*within* the octave from *P* to *2P*. These notes, from lowest to highest pitch, composed a Pythagorean *septatonic* scale—a set of seven notes with the following pitches:

$$\left\{P,\ \frac{9}{8}P,\ \frac{81}{64}P,\ \frac{4}{3}P,\ \frac{3}{2}P,\ \frac{27}{16}P,\ \frac{243}{128}P\right\}.$$

By adding the note corresponding to *2P*, we create a span of eight notes—hence, the term *octave*. The pitch of the perfect fifth relative to *P*—that is, $\frac{3}{2}P$—corresponds to the fifth note in this octave.

Although the structure of a musical scale is determined by the ratio of the frequencies of the consecutive notes in the scale, "the choice of these ratios is ultimately governed by the degree of consonance between the notes" (Bibby 2003, p. 13). The Pythagorean septatonic scale has the following consecutive ratios:

$$\frac{9}{8},\ \frac{9}{8},\ \frac{256}{243},\ \frac{9}{8},\ \frac{9}{8},\ \frac{9}{8},\ \text{and}\ \frac{256}{243}.$$

It is important to note that the Pythagoreans built their theory of consonant sounds on the notion that such sounds corresponded to string lengths that were commensurable. Hence, the consecutive ratios in their scale were all rational numbers.

The Pythagorean method is very efficient for producing particular scales but less efficient for transposing musical compositions from one key to another. The difficulty occurs especially in transposing the semitones between consecutive notes in the octave. Each octave in a musical scale is typically broken into twelve notes, collectively referred to as a chromatic scale. As a musician moves up the scale, each note that he or she plays is a semitone higher than its predecessor. The piano keyboard gives a good representation of this, as figure 3.1 shows. The number of black keys and white keys in any octave on the piano is twelve.

Students may be surprised to learn that the solution to the problem of transposing music to different keys involved introducing irrational numbers as the ratios of consecutive frequencies in a chromatic scale. This was not a solution that the Pythagorean scaling method allowed.

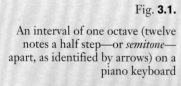

Fig. **3.1.**

An interval of one octave (twelve notes a half step—or *semitone*—apart, as identified by arrows) on a piano keyboard

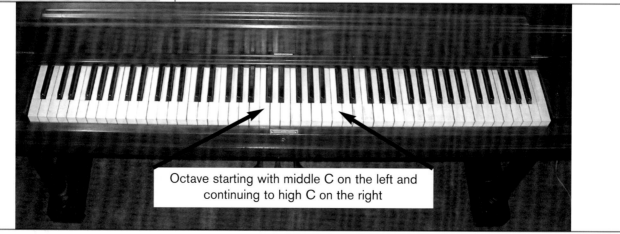

Octave starting with middle C on the left and continuing to high C on the right

One such scale is the well-tempered musical scale developed by Johann Sebastian Bach and others in the eighteenth century.

The activity Frequency, Scales, and Guitars guides students in investigating the properties of a particular kind of well-tempered scale—a twelve-note chromatic scale with equal space between consecutive notes in an octave. Students tend to think of irrational numbers as very abstract, without many practical purposes. This activity shows that this is not true of $\sqrt[12]{2}$, which emerges as a cornerstone of the equal-spacing method of determining musical scales. The students' work will help them gain insight into the use of irrational numbers in the construction of the guitar and other stringed instruments. Students can make a similar case for many other irrational numbers that they encounter in their mathematical work.

Some students in your class may be unacquainted with or confused by the ideas of a scale and an octave. Before beginning the activity, you may want to show a piano keyboard or a guitar and play several notes to help clarify these ideas. You can demonstrate that an octave is divided into 12 notes, each a semitone away from its predecessor or successor.

Before introducing the activity, you may also want to help your students visualize sound as a wave with a frequency, an amplitude, and a period. The activity does not probe this topic or the role that the tension on a guitar string plays in the pitch of a note produced on the string. The CD-ROM includes an applet activity, Sound Wave, which fills in these parts of the story.

After these preliminaries, your students should be prepared to begin Frequency, Scales, and Guitars. The activity opens with a review of some of the terms that describe sound waves, and the students apply the terminology in the computation of the frequency of a periodic graph in step 1. They analyze a graph of a sound wave produced by a tuning fork with a frequency rating of 440 hertz (cycles per second), corresponding to the musical note A. (Fig. 3.2 shows the graph that the students see.)

Well-tempered scales were important musical innovations of Johann Sebastian Bach (1685–1750) and others. A well-tempered scale is *equal-tempered* if it has equal spacing between the pitches of successive notes in an octave and uses $\sqrt[12]{2}$ to determine the frequencies of the notes.

 See "Roots in Music" (Houser 2002; available on the CD-ROM), or "Faggot's Fretful Fiasco" (Stewart 2003), for more detailed information on the idea of "well-tempering."

 The applet activity Sound Wave on the CD-ROM supplements the students' work in Frequency, Scales, and Guitars by linking a string's tension to its pitch and showing a visual model of the propagation of sound waves through air.

Note that the label for the *y*-axis is "Potential (volts)." An air wave causes a membrane in the microphone to move, and that membrane's motion is part of an electrical capacitance system and causes a fluctuation in the electrical potential of the capacitance system. Electrical potential is measured in volts.

A-440 Tuning Fork

(graph showing Potential (volts) vs Time (seconds), with labeled points (0.0098, 2.68817), (0.0121, 2.68817), (0.002, 2.65396), (0.0042, 2.65396))

Fig. **3.2.**

A graph of a sound wave produced by a tuning fork with a frequency of 440 Hz, corresponding to the note A

"In 1939 the International Standards Association, the same body that is responsible for S.I. units of measurement, agreed on a world-wide standard of 440 Hz for the note we now call *concert A*." (Johnston 2002, p. 36)

Concert A is the note that musicians use for tuning their instruments in an orchestra. It determines the frequencies of the other notes once the musicians decide on the "tempering" of the scale in use.

Most notes sounded by stringed or brass instruments are actually made up of several notes. The fundamental, or dominant, note has the lowest frequency of these notes. The other notes, called overtones, have higher frequencies that are generally multiples of the lowest frequency. See Johnston (2002, pp. 86–97) for an explanation of this phenomenon.

Tuning forks and oboes are examples of instruments that produce notes with minimal overtones. The purity of an oboe's sound may help explain why the tradition developed of having orchestra members tune their instruments to an A sounded by the principal oboe player.

As students work through step 1, they should realize that the graph is not smooth. The irregularity occurs because the instrument used to gather the data—a microphone—can collect only a relatively small number of data points in the short intervals of time that the graph represents. Thus, the data points that appear to correspond to the peaks and troughs of the wave do not necessarily correspond to its actual peaks and troughs. As a result, using different cycles in the graph to approximate the frequency of the sound wave gives significant variation.

One way to deal with the variation is to average; the students use this method in step 1. Figure 3.2 displays the coordinates that the activity sheet shows for two consecutive troughs and two consecutive peaks. The *x*-values (or times) at the peaks and troughs yield two estimates of the frequency. Using the times for the peaks, the students subtract to find one estimate of the period of the wave (that is, the time that elapses during one cycle of the wave):

0.0121 seconds – 0.0098 seconds = 0.0023 seconds.

Thus, the frequency of the wave is approximately $(.0023)^{-1}$, or 435, Hz. Using the times for the troughs, the students subtract again to find a second estimate of the period of the wave:

0.0042 seconds – 0.0020 seconds = 0.0022 seconds;

Fig. 3.3.

The bridge, nut, and frets on a guitar

thus, the frequency is approximately $(.0022)^{-1}$, or 455, Hz. Averaging these frequencies, the students obtain 444 hertz as an approximation of the frequency of the tuning fork.

The activity then explores scales and frequencies on a guitar, focusing on the instrument's A-string. Since an octave consists of twelve half steps, the A at the end of the string's first octave sounds when someone presses the string at the 12th fret and plucks or strums it. (See fig. 3.3 for the position of the 12th fret on a guitar.)

It is important to note that the sound from a guitar string is not as pure as the sound from a tuning fork. The graphs that the students examine in step 2 (see fig. 3.4) indicate the presence of *overtones* produced by an open vibrating A-string and a vibrating A-string that is pressed at the 12th fret. However, in spite of these overtones, the students can make reasonable estimates of the period and frequency of the notes. The students calculate and compare these values in step 2.

The frequency of the note produced by the guitar's open A string is about 110 hertz. The students should observe that the frequency of the 12th-fret A doubles to about 220 hertz.

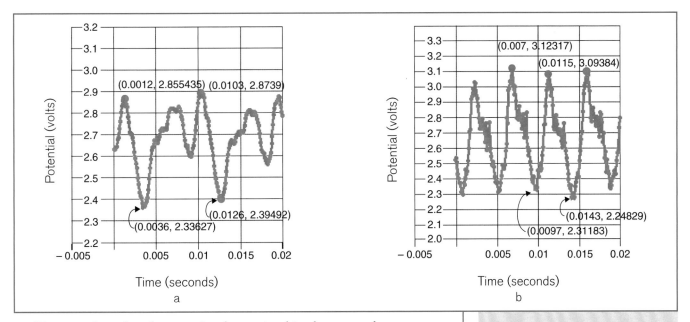

Fig. **3.4.**

Graphs of sound-wave data collected from a guitar's A-string, (a) open and (b) pressed at the 12th fret

How are the other frequencies determined in the octave between these two notes? Step 3 gives the students all the necessary frequency data in a table. To complete the table, they need only compute the differences and ratios between the frequencies given for successive notes. However, if you prefer, your students can use frequency data that they collect on their own with a microphone and a calculator-based laboratory (CBL). Either way, the students can explore the relationship between the fret number and the frequency of the corresponding note by analyzing patterns in the differences and ratios of consecutive frequencies in the scale. Table 3.1 shows the relationships for the given data.

The students discover that although the differences are not constant, the ratios of consecutive frequencies appear to be roughly 1.06. Thus, the frequency of the note sounded at the first fret is approximately 110×1.06, or about 117, hertz. The frequency of the note sounded at

Table 3.1

Relationships between Successive Frequencies for the First Twelve Notes of a Guitar's A-String

Fret	Note	Frequency (Hz)	Consecutive Differences (Hz) $(F_{n+1} - F_n)$	Consecutive Ratios $\left(\dfrac{F_{n+1}}{F_n}\right)$
0, or "open"	A	110	117 – 110 = 7	117 ÷ 110 = 1.06
1	A#	117	123 – 117 = 6	123 ÷ 117 = 1.06
2	B	123	8	1.06
3	C	131	8	1.06
4	C#	139	8	1.06
5	D	147	9	1.06
6	D#	156	9	1.06
7	E	165	10	1.06
8	F	175	10	1.06
9	F#	185	11	1.06
10	G	196	12	1.06
11	G#	208	12	1.06
12	A	220		

the second fret is approximately 110×1.06^2 hertz. Continuing this pattern, we find that the frequency of the note at the 12th fret is approximately 110×1.06^{12} hertz.

If we assume that the ratios of consecutive frequencies have a constant value r, then $220 = 110r^{12}$, since the note at the 12th fret has a frequency of 220 hertz, or double that of the note produced by the open A, which has a frequency of 110 hertz. Step 4 lets the students consider these ideas as they solve the equation $220 = 110r^{12}$ to discover that $r^{12} = 2$, and $r \sqrt[12]{2} \approx 1.06$.

This work in step 4 confirms the value that the students found empirically for the constant ratio in step 3. Moreover, it will reveal to the students that the sounds of the notes at consecutive frets on a guitar string form a geometric sequence of pitches—a geometric sequence that they can hear! To find the frequencies of the notes in the second octave of the A-string, as the students do in the last part of step 4, they simply multiply successive frequencies by $\sqrt[12]{2}$, starting with 220.

Steps 5–7 conclude the activity with an investigation of the physical configuration of the fretboard of a guitar. This exploration and analysis of the distances from the guitar's bridge to each of its first twelve frets leads the students once more to $\sqrt[12]{2}$. Again, the activity sheet provides the students with all the necessary measurement data, but you can have your students set these data aside and measure the distances themselves if you prefer. No matter which way your students proceed, they will look for a model that relates each fret number to its corresponding distance from the bridge. Table 3.2 shows the relationships between the successive distances for the given data. The table shows the differences and ratios that the students calculate with these data to complete the table in step 5.

After discovering the pattern in the earlier table (see table 3.1), the students should not be surprised to find that the string length from the bridge to the 12th fret is one-half the string length from the bridge to the

Table 3.2

Relationships between Successive Distances from the Bridge for the First Twelve Frets of a Guitar

Fret	Length (cm)	Consecutive Differences (cm) $(L_{n+1} - L_n)$	Consecutive Ratios $\left(\dfrac{L_{n+1}}{L_n}\right)$
0 (Nut)	64.8	–3.6	$61.2 \div 64.8 = 0.94$
1	61.2	–3.4	$57.8 \div 61.2 = 0.94$
2	57.8	–3.2	0.94
3	54.6	–3.2	0.94
4	51.4	–2.7	0.95
5	48.7	–2.8	0.94
6	45.9	–2.7	0.94
7	43.2	–2.3	0.95
8	40.9	–2.4	0.94
9	38.5	–2.2	0.94
10	36.3	–2.0	0.94
11	34.3	–1.9	0.94
12	32.4		

nut—the length of the open string—and that the ratios of consecutive lengths appear to be a constant, r. These discoveries, together with the information that the guitar's first fret is 64.8 centimeters from the bridge and its 12th fret is 32.4 centimeters from the bridge, yield the equation $32.4 = (64.8)r^{12}$, which the students consider and then solve in step 6:

$$32.4 = (64.8)r^{12}$$
$$r^{12} = \frac{1}{2}$$
$$r = \frac{1}{\sqrt[12]{2}} \approx 0.94.$$

They see that with this value for r, they can compute the locations of the frets on the guitar.

Thus, in the design of the guitar, as well as other stringed instruments, the reciprocal of $\sqrt[12]{2}$ is embedded in the spacing of the frets. The students should recognize that this spacing forms a geometric sequence of lengths—this time, a geometric sequence that they can see!

Moreover, since the frequencies and lengths are determined by the fret number, the students can relate the frequencies of the notes to the lengths (measuring from bridge to fret) of the guitar strings producing them, recognizing that the frequency of a note is inversely proportional to the string length. Step 7, the activity's last step, invites the students to reflect on and explain this relationship.

Assessment

The activity Frequency, Scales, and Guitars offers numerous possibilities for assessing your students' understanding. For example, to determine how well the students grasp the mathematics of the equal-tempered scale, you can ask them to do one of the following:

- Compute the frequencies of the notes in several octaves of the C-scale, beginning with middle C, which has a frequency of 264 hertz. (You can check to see whether your students base their answers on the function $y = 264\left(\sqrt[12]{2}\right)^x$, where y is the frequency and x is the fret number.)
- Measure the frets on various stringed instruments and determine models for the positions of the frets.
- Construct a fretboard and use it to play some songs.
- Use a data collection device with a microphone and determine the frequencies of the notes of a scale on a stringed instrument. Then compare these with an equal-tempered scale.

Another way to assess your students' understanding of the ideas of the activity is to have them investigate the Pythagorean method of constructing a musical scale. As the discussion indicates, the ancient Pythagoreans used ratios of only whole numbers to determine the pitches of notes. If we let w stand for a whole step and h stand for a half step, we can think of the eight main notes of an octave scale as the initial note of the scale followed by these steps: w, w, h, w, w, w, h. As we have seen, the Pythagorean method multiplies the frequency of a note by $\frac{9}{8}$ to produce each whole step and by $\frac{256}{243}$ to produce each half step.

Since $\frac{9}{8} \times \frac{9}{8} \times \frac{256}{243} \times \frac{9}{8} \times \frac{9}{8} \times \frac{9}{8} \times \frac{256}{243} = 2$, the Pythagorean method for determining the main notes in an octave guarantees that the frequency doubles from the first note to the first note of the next octave.

Table 3.3 juxtaposes frequencies for a classical scale resulting from the Pythagorean method with frequencies for an equal-tempered scale. For example, the frequency of the note E, denoted here as $F(E)$, is determined by

$$F(E) = F(A) \times 2^{\frac{7}{12}} = 220 \times 2^{\frac{7}{12}} \approx 330 \text{ Hz}$$

in the equal-tempered scale, and

$$F(E) = F(A) \times \frac{9}{8} \times \frac{9}{8} \times \frac{256}{243} \times \frac{9}{8} = 220 \times \frac{3}{2} \approx 330 \text{ Hz}$$

in the Pythagorean scale. In fact, $2^{\frac{7}{12}}$ and $\frac{9}{8} \times \frac{9}{8} \times \frac{256}{243} \times \frac{9}{8}$ are both very close to 1.5, making the two scales quite close at this point.

By investigating a classical scale based on the Pythagorean method, your students can gain a historical perspective on the significance of the

Table 3.3
A Comparison of a Classical Scale with an Equal-Tempered Scale

Note	Frequency with Pythagorean Ratios (Hz)	Frequency with Equal-Tempering (Hz)
A	220	220
B	$\frac{9}{8} \cdot 220 = 247.5$	247
C#	$\frac{9}{8} \cdot 247.5 = 278.4375$	277
D	$\frac{256}{243} \cdot 278.4375 = 293\frac{1}{3}$	297
E	$\frac{9}{8} \cdot 293\frac{1}{3} = 330$	330
F#	$\frac{9}{8} \cdot 330 = 371.25$	370
G#	$\frac{9}{8} \cdot 371.25 = 417.65625$	415
A	$\frac{256}{243} \cdot 417.65625 = 440$	440

changes instituted by J. S. Bach and others, who introduced equal-tempered and other well-tempered scales. Since the ratios of the frequencies of consecutive half steps in the Pythagorean scale are not uniform as in the equal-tempered scale studied in this activity, transposing a composition from one key to another causes disproportionate changes in the sound sequences and creates a significant difference in the melody. Well-tempering solves this problem and provides the basis for the design and tuning of many modern instruments, such as the guitar, mandolin, lute, and piano.

Where to Go Next in Instruction

The activity Frequency, Scales, and Guitars lets students discover the role of irrational numbers in the construction of well-tempered musical scales. Though the ancient Greeks relied on their intuitions about the sounds that they produced by using rational fractions of a string's length on a musical instrument, Bach and others tried to "standardize" the scale with the help of irrational numbers.

In the next activity, Rock Around the Clock, students continue to explore the pervasive use of numbers in the world around them. In this activity, they explore the construction of communication coding schemes that use prime numbers and finite number systems.

Rock Around the Clock

Goals

- Introduce students to clock arithmetic systems
- Study properties of integer multiplication modulo n
- Use multiplication modulo 31 for encoding and decoding a position cipher

Materials and Equipment

For each student—

- A copy of the activity sheet "Like Clockwork"
- A copy of the activity sheet "Encryption à la Mod"
- A copy of the activity sheet "Ciphering in Mod 31"
- A copy of the activity sheet "Make a Code / Break a Code"
- A calculator

Discussion

Codes and the secrets that they conceal have a perpetual fascination. Mathematics teachers can take advantage of students' curiosity about codes to explore the interesting mathematics of cryptography.

Cryptographers call the simplest codes *substitution ciphers*. In these ciphers, each punctuation mark, numeral, or letter of the alphabet takes the place of a different letter, numeral, or punctuation mark. For example, using the substitution values in figure 3.5, a cryptographer could encode the message *Wheel-of-Fortune* as *ETWWPAMVAVMJHGNW*. Substitution ciphers are easy to "crack," or decode, because of the relative frequency with which particular letters appear in messages. Fans of the popular TV game show *Wheel of Fortune* know the most common letters in words and phrases. For example, *E* is the most frequently occurring letter in the English language. Thus, the substituted value for an *E* is usually relatively easy to determine in a simple substitution cipher.

Fig. **3.5.**

A simple substitution cipher for plaintext

A	B	C	D	E	F	G	H	I	J	K	L	M	N	O
?	Z	Y	X	W	V	U	T	S	R	Q	P	O	N	M
P	Q	R	S	T	U	V	W	X	Y	Z	-	.	,	?
L	K	J	I	H	G	F	E	24	C	B	A	,	.	30

One method of making substitution codes more difficult to decipher is to use a coding scheme that depends on the position of the characters in a message. Such a code is called a *position cipher*. The activity Rock Around the Clock presents a particular application of this strategy. This coding scheme combines the ideas of prime number, multiplication, modular arithmetic, and inverses to render *Wheel-of-Fortune* as *WPOT,GLQZ,J?LOXR*. Note that the *E*s that occur in *Wheel-of-Fortune* do not correspond to the same letter in the encoded message. In this position cipher, their encoding depends on their position in the original phrase.

Rock Around the Clock consists of four parts that take students step by step from the basics of modular arithmetic—also known as "clock arithmetic"—to the use of mod 31 for encoding and decoding simple messages. In part 1, students explore the idea of "walking" a number around a "clock." In part 2, they investigate properties of multiplication modulo 7. They extend these properties to mod 31 in part 3. In part 4, they apply multiplication modulo 31 to actual tasks of ciphering and deciphering messages.

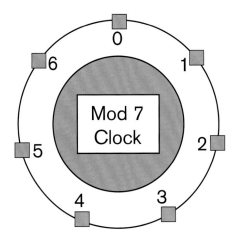

Part 1—"Like Clockwork"

After the students learn what it means in modular arithmetic to "walk" a number around a clock, they walk the number 20 around a mod 7 clock in a clockwise direction and look for patterns in the numbers that "stop" in the same spot on the clock. If they walked 3925 around the clock, where would it stop? They consider possible shortcuts for determining the answer to this question.

They should notice that a number's position on the clock is the same as the remainder that they obtain when they divide the number by 7. For example, $20 \div 7$ has a remainder of 6. Hence, when they walk 20 around the clock starting at 0, they stop at 6. The students learn that the notation $[n]_7$ refers to the remainder when n is divided by 7. Thus, $[20]_7 = 6$.

The students evaluate $[39]_7$, $[7 \times 69 + 2]_7$, $[7 \times 439 + 39]_7$, and $[7n + 3]_7$, gradually realizing that for integers h and k both greater than or equal to 0, $[7h + k]_7 = [k]_7$. The stop that they found for 20 on the mod 7 clock supports this conclusion: $20 = 7(2) + 6$, so the stopping position of 20 is 6.

The students then consider the division algorithm for integers in relation to the problem of determining stopping positions on the mod 7 clock. This algorithm ensures that for any positive integer p, unique integers q and r exist such that $p = 7q + r$, where $0 \leq r < 7$. The division algorithm should be no mystery to students. It summarizes what they already know about long division of positive integers—that the long-division process yields a unique quotient and remainder, and the remainder is less than the divisor but greater than or equal to 0. This fact has theoretical importance in part 2, where several proofs involve the assumption that every positive integer n can be expressed uniquely in the form $n = 7q + r$, where q and r are integers and $0 \leq r < 7$.

Once your students have grasped the connection between stopping positions on the mod 7 clock and division by 7, they will be ready to look for shortcuts to solving multiplication problems. They should discover that they can determine the remainder of a product like $[29 \times 47]_7$ by noting that $[29]_7 = 1$ and $[47]_7 = 5$ and $[1 \times 5]_7 = [5]_7 = 5$. Thus, $[29 \times 47]_7 = 5$. They should see, in other words, that in a product $n \times m$, they can replace a factor (n or m or both) with its remainder modulo 7 without affecting the outcome—the remainder of $n \times m$ when divided by 7.

The students prove this proposition and reflect on its significance in the last steps of part 1. The proof involves expressing both n and m in the form $7a + b$, where a and b are integers, as guaranteed by the division algorithm. The students let $n = 7q + r$ and $m = 7p + s$. They substitute first for n in the product $[n \times m]_7$:

$$[n \times m]_7 = [(7q + r) \times m]_7 = [7qm + rm]_7.$$

The division algorithm for integers claims that if n and $d \geq 1$ are integers, then there exist uniquely determined integers q and r such that $n = dq + r$ and $0 \leq r < d$.

Chapter 3: Number and Operations in the World

Then, using the fact that for any integers h and k, $[7h + k]_7 = [k]_7$, they get $[7qm + rm]_7 = [r \times m]_7$. Therefore, replacing n with r, or its remainder modulo 7, does not change the value of $[n \times m]_7$. Likewise, when the students substitute $7p + s$ for m and repeat the argument, they get

$$[r \times m]_7 = [r \times (7p + s)]_7 = [7pr + rs]_7 = [r \times s]_7.$$

Thus, replacing m with s, or its remainder modulo 7, also has no effect on the value of $[n \times m]_7$.

Part 2—"Encryption à la Mod"

In part two, the students investigate the mod 7 encryption numbers (1, 2, 3, 4, 5, and 6) under the mod 7 multiplication operation, defined as $n \otimes m = [n \times m]_7$. The students complete an operation table for mod 7 (see fig. 3.6), which reveals many important facts about this system of number and operation.

Fig. 3.6.
Multiplication modulo 7

$\otimes$	1	2	3	4	5	6
1	1	2	3	4	5	6
2	2	4	6	1	3	5
3	3	6	2	5	1	4
4	4	1	5	2	6	3
5	5	3	1	6	4	2
6	6	5	4	3	2	1

The students discover that the system is closed under the operation $\otimes$ and has an identity element. They find that each number in the system has a multiplicative inverse, and the operation $\otimes$ is commutative. Using what they know about remainders of numbers divided by 7, the students explain why the system has these properties.

In the language of abstract algebra, the system forms an *abelian group*. The one property of such a group that is not evident from the table is associativity. The activity asks the students whether or not the operation $\otimes$ is associative. Does $a \otimes (b \otimes c) = (a \otimes b) \otimes c$ for any mod 7 encryption numbers a, b, and c?

Your students may base their conclusion on random tests of specific cases, and you should allow them to do so. However, they may see that a proof of the associativity of $\otimes$ follows from their discovery in part 1 that in a product $n \times m$, replacing a factor (n or m or both) with its remainder modulo 7 does not affect the outcome of finding the remainder of $n \times m$ when divided by 7.

From this work in part 1, the students know that $[a \times (b \times c)]_7 = [a \times [b \times c]_7]_7$, which by definition equals $a \otimes (b \otimes c)$. Likewise, they know that $[(a \times b) \times c]_7 = [[a \times b]_7 \times c]_7$, which by definition equals $(a \otimes b) \otimes c$. Since integer multiplication is associative—that is, $a \times (b \times c) = (a \times b) \times c$—they can conclude that $[a \times (b \times c)]_7 = [(a \times b) \times c]_7$ and, therefore, $a \otimes (b \otimes c) = (a \otimes b) \otimes c$.

Students who have access to programmable calculators can easily check relatively small finite systems for associativity. Figure 3.7 shows one program that they can use. Written for a TI-73 calculator, the program checks every possible case in the system to see if associativity holds. For the mod 7 system, this involves checking 6^3, or 216, cases.

On a TI-73, the command **Remainder(n,7)** gives the remainder when an integer n is divided by 7. For the integers L, M, and N greater than or equal to 0 and less than or equal to 6, the program checks every possible case to see if it is ever true that $L \otimes (M \otimes N) \neq (L \otimes M) \otimes N$. If the program finds such a case, it returns "**Not Associative**" and the values of L, M, and N for which this is true. If it fails to find such a case, it returns "**Associative**."

Students can easily modify the program in figure 3.7 for other modular systems and other calculators. To check associativity for the mod 31 system, for example, they would need only replace each 6 in the program by 30 and each 7 by 31. To check associativity for a modular system on a calculator other than a TI-73, students may have to use slightly different commands. On a TI-83 Plus graphing calculator, for instance, checking associativity for mod 7 would require the command **n−7×int(n/7)**, which gives the remainder of dividing an integer n by 7. Run-time for such checks can vary from calculator to calculator and from one mod system to another. To check the mod 31 system for associativity, some calculators might take several minutes because they would have to check 30^3, or 27,000, cases.

Part 2 also guides the students in exploring the properties of the mod 6 system and comparing them with those of the mod 7 system. This work enables the students to discover the key to the usefulness of prime numbers in a position system of encryption. If n is a prime number, the set of integers $\{1, 2, 3, ..., n-1\}$ forms a group under multiplication mod n. This means that each number $1, 2, 3, ..., n-1$ has a multiplicative inverse modulo n. When n is a composite positive integer, this is not the case. In the mod 6 system, for example, as the students discover, only 1 and 5 have multiplicative inverses. These are exactly the integers that are relatively prime to 6 under normal integer multiplication.

There is one final step that the students must take before learning the method that the activity presents for coding and decoding position ciphers. They must create a table of inverses for the integers $1, 2, 3, ..., 30$ under multiplication mod 31. If your students need assistance, you can offer an effective strategy. Tell them to look for pairs of integers between 1 and 30 whose products are one more than a multiple of 31. For example, $63 = 31(2) + 1$. Therefore, 7 and 9 are multiplicative inverses mod 31. But since $-7 \times -9 = 63$, the numbers on the mod 31 clock corresponding to -7 and -9 are also multiplicative inverses modulo 31. Since $31 - 7 = 24$, and $31 - 9 = 22$, then 24 and 22 are also inverses modulo 31. Figure 3.8 shows all the multiplicative inverses mod 31 for the integers from 1 to 30.

TI-73 Program

PROGRAM: ASSOCIATE
:For(L,1,6)
:For(M,1,6)
:For(N,1,6)
:If remainder(L*remainder(M*N,7),7)
≠
remainder(remainder(L*M,7)*N,7)
:Then
:Disp L,M,N
:Disp "NOT ASSOCIATIVE"
:Stop
:End
:End
:End
:End
:Disp "ASSOCIATIVE"
:Pause

Fig. **3.7.**

A program on a TI-73 calculator for testing mod 7 multiplication for associativity

x	1	2	3	4	5	6	7	8	9	10	11	12	13	14	15
x^{-1}	1	16	21	8	25	26	9	4	7	28	17	13	12	20	29
x	16	17	18	19	20	21	22	23	24	25	26	27	28	29	30
x^{-1}	2	11	19	18	14	3	24	27	22	5	6	23	10	15	30

Fig. 3.8.
Multiplicative inverses in the mod 31 system

Part 3—"Ciphering in Mod 31"

Part 3 takes the students through the steps of using a position cipher to encode the message *I have a secret*. To encode or decode with a position cipher, a cryptographer needs to create a list of substitution values for each alphabetic character and punctuation mark. The particular technique that the students learn in Rock Around the Clock also requires that the number of substitution values be one less than a prime number. Since there are twenty-six letters in the English alphabet, the prime number 31 allows substitution values for all the alphabetic characters plus four punctuation marks. The character for the value 27 is "–", a short dash. This position system uses this character to represent the space between words.

Figure 3.9 shows the set of thirty substitution values that the activity sheet presents for these characters. Needless to say, many other substitution schemes are possible. In fact, there are 30! ways of assigning the numbers 1 through 30 to the characters in the table.

As the students know from the last step of part 2, the numbers 1 through 30 in figure 3.9 all have inverses under mod 31 multiplication. For example, $7 \otimes 9 = [7 \times 9]_{31} = [63]_{31} = 1$, a product that indicates that 7 and 9 are inverses under mod 31 multiplication. There is no character with a value of 31 since $[31]_{31}$ is 0, and 0 does not have an inverse under mod 31 multiplication.

To encode their message, the students complete a table, filling in the characters of the message and the corresponding position values, substitution values, ciphertext values, and ciphertext for the encoded message. Figure 3.10 shows the completed table.

To fill in row 1, the students simply enter the plaintext of the message letter by letter. They must remember that the message ends in a period, and they must be sure to use a dash for the space between words.

Completing row 2 is also easy—the first character in the plaintext has a position value of 1, the second character has a position value of 2, and so on. (Students learn that for a message of more than thirty characters, the counting scheme would begin again at 1 with the thirty-first character.)

Supplying the values in row 3 is also straightforward. To fill in this row, the students use the substitution values for plaintext (shown here as fig. 3.9).

Fig. 3.9.
Substitution values for plaintext

A	B	C	D	E	F	G	H	I	J	K	L	M	N	O
1	2	3	4	5	6	7	8	9	10	11	12	13	14	15
P	Q	R	S	T	U	V	W	X	Y	Z	–	.	,	?
16	17	18	19	20	21	22	23	24	25	26	27	28	29	30

Plaintext	I	–	H	A	V	E	–	A	–	S	E	C	R	E	T	.
Position values	1	2	3	4	5	6	7	8	9	10	11	12	13	14	15	16
Substitution values	9	27	8	1	22	5	27	1	27	19	5	3	18	5	20	28
Ciphertext values	9	23	24	4	17	30	3	8	26	4	24	5	17	8	21	14
Ciphertext	I	W	X	D	Q	?	C	H	Z	D	X	E	Q	H	U	N

Fig. 3.10.

A completed table for encoding the message *I have a secret.*

The task of completing row 4 is a bit more complex. The students discover that they must use modular multiplication to complete this row. They must determine the product mod 31 of each character's position value and its substitution value.

Completing row 5 is again straightforward. The students simply use the substitution values, as in row 3.

If you think your students will have difficulty following these steps on their own, you can conduct this part of the activity as a group lesson in encoding. For this lesson, you can make transparencies from the blackline activity pages and go through them step by step with your students.

Part 4—"Make a Code / Break a Code"

The students are now set to apply what they have learned about using mod 31 multiplication to the creation of their own position ciphers. In part 4, they complete tables like that in figure 3.10 to encode two new messages: *Math is useful* and their first names.

This part of the activity also introduces the students to the process of decoding a message. They consider two messages that someone has encoded with a position cipher and mod 31 multiplication. Now the students discover the usefulness of the multiplicative inverses of encryption numbers. They again consider the message *I have a secret.* When they encrypted this message in part 3, they encoded each character by multiplying its position value by its substitution value. To decode the resulting message *IWXDQ?CHZDXEQHUN*, they must multiply each ciphertext value by the multiplicative inverse of its position value.

The students thus see that a table for decoding such a message needs one more row than a table for encoding it, since it must include a row for the inverses of the position values. Figure 3.11 shows the completed

Fig. 3.11.

A completed table for decoding the message *IWXDQ?CHZDXEQHUN*

Ciphertext	I	W	X	D	Q	?	C	H	Z	D	X	E	Q	H	U	N
Position values	1	2	3	4	5	6	7	8	9	10	11	12	13	14	15	16
Inverses of position values	1	16	21	8	25	26	9	4	7	28	17	13	12	20	29	2
Substitution values	9	23	24	4	17	30	3	8	26	4	24	5	17	8	21	14
Plaintext values	9	27	8	1	22	5	27	1	27	19	5	3	18	5	20	28
Plaintext	I	–	H	A	V	E	–	A	–	S	E	C	R	E	T	.

table for decoding the message *IWXDQ?CHZDXEQHUN*. The students examine this table and then complete such a table to decode a new message on their own.

You might challenge your students to use calculators or electronic spreadsheets to facilitate the coding and decoding processes, though by-hand calculations are adequate for the uncomplicated exercises in the activity. You can also encourage your students to explore extensions and modifications of the method that the activity presents. Again, if your students are likely to find the process of decoding excessively difficult to master on their own, you can make transparencies from the blackline pages for part 4 and use them to facilitate a whole-class lesson.

Assessment

One of the challenging aspects of this activity is mastering new notations and mathematical systems. It is important to assess how well the students understand the notations $[m]_7$ and $p \otimes q$. If they can explain these notations in terms of the numbers on the mod 7 clock, then they will be prepared to explore the mod 7 system. And if they can keep in mind the connection between multiplication mod n and division by n for integers, they can make sense of the problems that the activity poses.

Some students will want to use calculators to find the remainders of divisions by 7. Allowing them to do so will give you a good opportunity to assess their understanding of the division operation for integers. For example, you might have students who divide a number such as 405 by 7 on their calculators, obtain a value of 57.85714286, and interpret the remainder as .85714286 instead of as 6, as they need to in mod 7. You should make sure that your students can perform the operation with a by-hand algorithm and can interpret the results of their calculators in light of the paper-and-pencil approach.

Many extensions can reveal students' understanding of the activity's main concepts. For example, you might ask your students to study position ciphers that use the mod 29 system, limiting punctuation marks in their messages to the dash and the period. Alternatively, you can have them choose a different modular number system and devise their own coding schemes, perhaps by limiting the alphabetic characters that they use. For a more theoretical extension, ask your students to investigate the addition operation on the mod 7 clock.

Exploring cryptography—a robust and growing branch of modern mathematics—can offer students a chance to develop a deeper understanding of number and operations. As this activity illustrates, the study of encoding and decoding secret messages can give students important experience in using number systems to solve realistic problems.

Conclusion

The activities in chapter 3 have illustrated some of the rich possibilities for extending students' understanding of number and operation by investigating the mathematics embedded in sophisticated instruments and processes in the real world. Using technology with these and other activities can enhance students' appreciation and enjoyment of the mathematics while helping them learn important skills and concepts.

Additional information about position ciphers appears on the CD-ROM. "Calculator Cryptography" (Hall 2003) and "Exploring Hill Ciphers with Graphing Calculators" (St. John 1998) both make use of matrix multiplication and modular arithmetic for coding and decoding messages.

Chapter 4, the book's final chapter, focuses on rewarding problem-solving situations that can serve as springboards for multiple, interrelated studies of number and operations. These problem-solving contexts call into play all five aspects of learning that the Process Standards address. The situations in the problems highlight the importance of the five "process" elements that *Principles and Standards* enumerates: problem solving, reasoning and proof, communication, connections, and representation.

NAVIGATIONS SERIES

GRADES 9–12

NAVIGATING through NUMBER and OPERATIONS

Chapter 4
Extending Number and Operations Activities

"Working such problems [as 'what can you say about the number that results when you subtract 1 from the square of an odd integer?'] deepens students' understanding of number while providing practice in symbolic representation, reasoning, and proof."
(NCTM 2000, p. 292)

Principles and Standards for School Mathematics (NCTM 2000) recommends offering students in grades 9–12 varied opportunities to apply mathematics that they already know to the study of number. Teachers create such opportunities when they ask their students questions like the sample that *Principles and Standards* proposes: "What can you say about the number that results when you subtract 1 from the square of an odd integer?" (p. 292). This question represents a simple problem that students can address with algebraic reasoning.

With such number problems, students can quickly generate examples to support or refute conjectures. Problems of this type lend themselves to multiple approaches and create good contexts for discussing the differences between inductive evidence and deductive proof. Moreover, as this chapter illustrates, such problems can be springboards to progressively more profound mathematical investigations—especially in classrooms that encourage students to make full use of the elements of the Process Standards—problem solving, reasoning and proof, communication, connections, and representation.

This chapter introduces several number problems and illustrates how to elaborate them, stage by stage, into deeper, increasingly sophisticated investigations. Three activities show the riches that you and your high school students can discover in seemingly elementary problems.

The first activity, Number Triangles, provides a simple context for introducing systems of linear equations. It also demonstrates the power of variables as tools for proving conjectures for which the students have only empirical evidence. Number Triangles has three parts that students can work as a single exploration or as three separate explorations.

The second activity, Perfect Squares, starts with an investigation of the basic properties of perfect squares and leads students to an investigation of a finite number system with some interesting properties. In this activity, as in Number Triangles, conjecture, reasoning, and proof emerge naturally as students go beyond the empirical evidence provided by their calculations. Furthermore, the context is so rich that connections and multiple representations mesh effectively with the problem-solving process to create a set of interrelated activities that you can easily expand if you wish.

The third activity, Flooding a Water World, is a counting activity. The students explore simple connected networks of towers and dikes in an imaginary world covered by water. They discover the relationships among the numbers of towers, dikes, and cantons—habitable regions surrounded by towers and dikes. By working together, comparing networks, and verifying findings in this relatively simple context, they gradually gather information that they can apply in other, more sophisticated geometrical contexts.

Number Triangles

Goals

- Apply an understanding of the properties of whole numbers in solving a number problem
- Use systems of equations to discover properties of solutions to a number problem
- Prove that an iterative solution strategy is valid for a number problem

Material and Equipment

For each student—

- A copy of the activity sheet "Probing the Pattern"
- A copy of the activity sheet "It All Adds Up"
- A copy of the activity sheet "Take a Trip around a Triangle"
- A calculator

pp. 124–26; 127–28; 129–31

Discussion

A *number triangle* is a triangular arrangement of six integers as shown in figure 4.1. The integers are arranged on each side of the triangle so that x, y, and z are all sums of the numbers at the two adjacent vertices. Hence, $x = a + b$; $y = b + c$; and $z = c + a$.

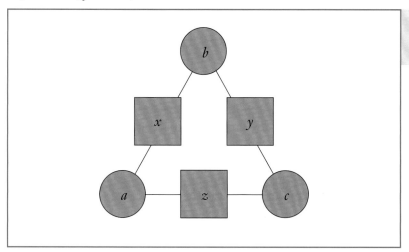

Fig. **4.1.**
A number triangle

The activity Number Triangles has three parts. The students explore patterns in such triangles in part 1. In part 2, they probe solutions and the characteristics of number triangles that have no solution. Part 3 lets the students explore and verify a particular solution strategy.

Part 1—"Probing the Pattern"

The students begin their work in part 1 by discovering the rule for the arrangement of the numbers in the triangle. Most students will quickly note that the numbers on the sides of the triangle are the sums of the numbers at the adjacent vertices. However, to be sure that your students see the correct pattern, you should have them compare their answers as a class or in groups.

The students go on to consider possible arrangements of even and odd numbers in a number triangle, and then they do the same for positive and negative numbers. They apply their discoveries as they complete two number triangles. You should encourage your students to find different ways to solve the problems. Many are likely to use guess and check, but others will probably use algebra to reason their way to a solution, and some may use a combination of the two.

When solving a number triangle such as that shown in figure 4.2, students can use algebra informally to narrow the possible values for a, b, and c. For example, students might note that $\frac{27 + 55 + 46}{2}$ is the sum $a + b + c$, or 64. Alternatively, by studying patterns in completed number triangles such as those in step 1, a student might notice that the difference between y and x always equals the difference between c and a. They can confirm this algebraically: $x - y = (a + b) - (c + b) = a - c$. With this information, the students can readily solve the triangle: $c - a = 55 - 27 = 28$. But $c + a = 46$. With simple algebra or guess and check, they easily discover that $c = 37$ and $a = 9$.

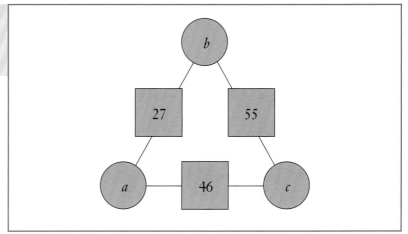

Fig. **4.2.**

A number triangle problem to solve by finding values for *a*, *b*, and *c*

It is important for your students to share and discuss their solution strategies. In these exchanges, the students will find that they formulate different conjectures about number triangles. For example, some students may notice that the numbers in the three squares cannot all be odd numbers. Others may observe that the sum of the three numbers in the squares cannot be odd but in fact must be twice the sum of the three numbers in the circles.

Such conjectures can give you a good context in which to focus on reasoning and proof. These two conjectures are of course related, as the students who make them should realize by proving the second conjecture. The first conjecture will follow automatically as a corollary.

Part 2—"It All Adds Up"

In part 2 of the activity, the students focus on number triangle problems as a class, and they generalize the conditions under which solutions exist. As elsewhere in the activity, the students must keep in mind that the numbers in the triangles are restricted to the integers — and sometimes specific subsets of the integers. Use part 2 to create open-ended dialogues about the validity of various conjectures and ways of proving or disproving them.

Part 3—"Take a Trip around a Triangle"

In part 3 of the activity, the students represent number triangle problems as systems of equations. Some of your students may have done this already in part 1. However, part 3 explicitly invites the students to use systems of equations to solidify discoveries from parts 1 and 2 as well as to prove other general results about number triangles.

For example, step 3 asks the students to prove that in a number triangle such as that in fig. 4.1, no integers a, b, and c exist if $x + y + z$ is an odd number. Students might add all three equations in the system, $a + b = x$, $b + c = y$, and $c + a = z$, to obtain the equation $2(a + b + c) = x + y + z$. Because the values of a, b, and c and x, y, and z are restricted to integers, the students know from this equation that $x + y + z$ must be even. In fact, by solving the general system of equations, the students can discover that $a = \frac{x + z - y}{2}$, $b = \frac{x + y - z}{2}$, and $c = \frac{y + z - x}{2}$.

Thus, the students know that there are always integer solutions for number triangles as long as $x + y + z$ is even. (If $x + y + z$ is even, then so are $x + y - z$, $x + z - y$, and $y + z - x$.)

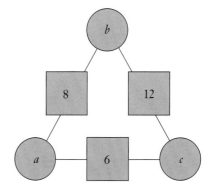

The students' work with number triangles culminates in an investigation of an iterative guessing method for solving number triangles. The students suppose that in another classroom a student solves the number triangle shown in the margin by the following method:

> "Say that I guess that a = 4. Because b = 8 – a, I have to say that b = 4. I also know that c = 12 – b, so c has to be 8. I know too that a = 6 – c, which gives me a = –2. But I started with a = 4. OK. So how about if I split the difference between 4 and –2, or average them? This gives me a = 1. If I use this value for a, I force b to be 7 and c to be 5. Look! Now I have a solution that works!"

The method is very efficient and reliably yields a solution as long as $x + y + z$ is even. The challenge to students is to prove that the method works. Once again, the students discover that algebraic reasoning is the key to proving the result.

Assessment

The activity Number Triangles restricts the numbers in the triangle to integers. Despite this restriction, students generally can solve number triangles after a little thought. However, they encounter bigger challenges when they must reason about the possible solution sets for whole classes of number triangles.

For example, in part 1, "Probing the Pattern," steps 3 and 4 ask the students to consider possible arrangements of odd numbers and negative numbers in a number triangle. These steps can give you opportunities to assess your students' ability to reason about the properties of odd and even numbers and of positive and negative numbers in the context of number triangles. Your students should be able to see that all six numbers in a number triangle cannot be odd. The students should be able to explain why this and other such facts are true.

Expanding the solution set first to rational numbers and then to real numbers can give you a chance to assess your students' understanding of relationships in number triangles. The students' work under the new conditions can show you how well they can engage in numeric

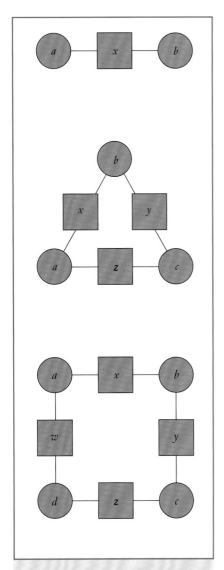

Fig. **4.3.**

Three in a series of related problems

reasoning and grasp the essential differences between number systems. Your students should be able to discover and prove that in a number triangle of the form in figure 4.1, whenever x, y, and z are rational or real numbers, a unique solution exists.

To create other activities that can allow you to assess your students' understanding of number triangle problems, you might restrict the numbers to specific sets of integers. For example, if in a number triangle of the form in figure 4.1, x, y, and z are multiples of 3, must a, b, and c also be multiples of 3? (The students should discover that they must.)

You can also assess your students' understanding of the main concepts in the activity Number Triangles by introducing other types of number arrangements based on the same rule—that an integer in a square is the sum of integers in the two adjacent circles. For example, your students could consider the series of similar number problems in figure 4.3. You could pose a variety of questions about these problems:

- "What methods can you use to solve these problems?"
- "What are the possible solution sets for these problems?"
- "What patterns can you find in one or more of the problems?"

Where to Go Next in Instruction

Number Triangles offers students a rich problem-solving context in which to use number and operation for conjecturing, reasoning, and proof. It starts with a basic number and operation problem and progresses to the more challenging issues of solvability and determining the effectiveness of algorithms.

The next activity, Perfect Squares, provides another context that reveals greater depth and richer rewards as the students continue to explore it. This activity introduces problems that lead students step by step from a focus on standard integer operations to reasoning and proof in a new number system—the integers modulo 10.

Perfect Squares

Goals

- To study properties of a class of numbers with special importance in number theory
- To solve problems by using the properties of perfect squares and last digits
- To investigate a finite number system created from the integers

Materials and Equipment

For each student—

- A copy of the activity sheet "Fair and Square"
- A copy of the activity sheet "Last-Digit Revelations"
- A copy of the activity sheet "The Singles Club"
- A calculator

pp. 132–34; 135–38; 139–40

Discussion

The study of triangular, square, pentagonal, and other *polygonal numbers* through their representations in geometric figures (see fig. 4.4) dates back to antiquity. As the history of mathematics demonstrates, examination of any of these classes of numbers can be a starting point for numerous excursions into number and operations. Likewise, classroom examinations of such numbers can transport students to investigations that will deepen their understanding of important ideas. This activity illustrates this process in an exploration of the square numbers, or *perfect squares*.

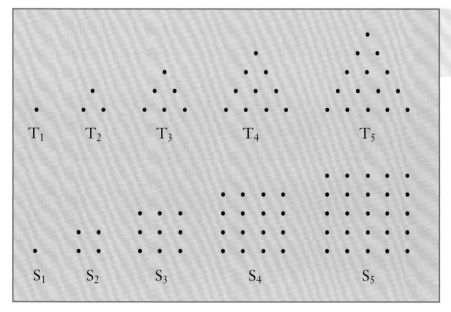

Fig. **4.4.**
Triangular and square numbers

Studies of perfect squares abound in the history of mathematics. Familiar examples in elementary mathematics include studies of Pythagorean triples (integers x, y, and z such that $x^2 + y^2 = z^2$), studies of

Chapter 4: Extending Number and Operations Activities   67

the sum of perfect squares $\left(1^2 + 2^2 + 3^2 + \ldots + n^2 = \dfrac{n(2n+1)(n+1)}{6}\right)$, studies of the sum of reciprocals of the perfect squares $\left(\dfrac{1}{1^2} + \dfrac{1}{2^2} + \dfrac{1}{3^2} + \ldots = \dfrac{\pi^2}{6}\right)$, and studies of primes. (Every prime of the form $4n + 1$ is the sum of two perfect squares. No prime of the form $4n - 1$ is the sum of two perfect squares. Whether an infinite number of primes exist that are one greater than a perfect square—that is, are of the form $n^2 + 1$—is unknown.)

Many investigations of perfect squares are accessible to students in grades 9–12—especially when students can use calculators. The activity Perfect Squares illustrates how you can apply the Process Standards to the study of number and operation in interrelated activities that start with a basic investigation of a class of numbers and end with the construction of a system of numbers and operations.

Many properties distinguish perfect squares from other integers. Students will have already encountered one very interesting property if they have completed the activity When Fractions Are Whole in chapter 1. In that activity, the students establish the fact that if a positive integer is not a perfect square, then its square root is irrational. In the three parts of the current activity, the students add other properties to the list.

Part 1—Fair and Square

Part 1 focuses on some of the basic properties of perfect squares that distinguish them from other integers. The students begin by considering the definition of *perfect square* and sorting a small group of numbers into those that are perfect squares and those that are not.

The students go on to explore patterns in the divisors and prime factorizations of perfect squares (see table 4.1). Unlike other positive

> *A perfect square is a positive integer w such that $w = n^2$ for some positive integer n. Thus, a perfect square has a positive integer as a square root.*

Table 4.1
Positive Divisors and Prime Factorizations of Integers That Are Perfect Squares and Integers That Are Not Perfect Squares

First Ten Perfect Squares	Number of Positive Divisors	Prime Factorization in Exponential Form	First Ten Positive Integers That Are Not Perfect Squares	Number of Positive Divisors	Prime Factorization in Exponential Form
1	1	1	2	2	2
4	3	2^2	3	2	3
9	3	3^2	5	2	5
16	5	2^4	6	4	2×3
25	3	5^2	7	2	7
36	9	$2^2 \times 3^2$	8	4	2^3
49	3	7^2	10	4	2×5
64	7	2^6	11	2	11
81	5	3^4	12	6	$2^2 \times 3$
100	9	$2^2 \times 5^2$	13	2	13

integers, a perfect square has an odd number of divisors. This is so because divisors of a positive integer come in pairs, with one divisor larger and the other divisor smaller than the square root of the number—unless the square root of the integer is itself an integer. In this case, the positive integer is a perfect square. For example,

$$6 = 6 \times 1 = 3 \times 2 \text{ and } \sqrt{6} \approx 2.4495;$$

thus, 6 has four positive divisors. However,

$$16 = 16 \times 1 = 8 \times 2 = 4 \times 4;$$

thus, 16 has five positive divisors, the "extra" divisor being the square root of 16.

The exponents in the prime factorization of a perfect square are always even numbers. Your students can easily demonstrate this fact by using the laws of exponents. If n and m are positive integers and $n = m^2$, then the prime factorization of n is the square of the prime factorization of m, making all of its prime-factor exponents even. For example,

$$144 = 12^2 = \left(2^2 \times 3^1\right)^2 = 2^{2 \times 2} \times 3^{1 \times 2} = 2^4 \times 3^2.$$

Part 1 concludes with an investigation of the properties of the differences between two perfect squares. By looking at the patterns in differences, students discover that every odd number is the difference between two consecutive perfect squares. For example,

$$3 = 2^2 - 1^2, 5 = 3^2 - 2^2, \text{ and } 7 = 4^2 - 3^2.$$

The students must use algebra to show that this pattern continues. They can do this by showing that if $(n-1)^2$ and n^2 are two consecutive squares, then their difference, $n^2 - (n-1)^2$, is $n^2 - (n^2 - 2n + 1)$, or $2n - 1$, which is the nth odd number. Therefore, every odd number can be the difference between two squares

In a similar fashion, the students can use the basic algebraic identity $x^2 - y^2 = (x - y)(x + y)$ to show that for the even integers, only the multiples of 4 are differences between two squares. The students can demonstrate that this is so because if n is even and $n = x^2 - y^2$, then one of the factors of n, $(x - y)$ or $(x + y)$, must be even. But they know that if one of these factors is even, then so is the other. Hence, n is the product of two even numbers, or, equivalently, a multiple of 4.

Part 2—"Last-Digit Revelations"

Part 2 takes the students from a consideration of the limitations of technology in representing large perfect squares to an examination of the last digits in perfect squares. The students consider the patterns in these digits and begin to understand their usefulness in identifying and probing large perfect squares.

Most calculators round off computations to less than fourteen significant digits. Thus, these calculators cannot give exact expressions for the squares of such large numbers as 23,456,789, the example on the activity sheet. (The TI-89, TI-92 Plus, and Voyage 200 calculators are notable exceptions, since they allow integer operations to compute up to 614 significant digits.)

You could challenge your students to develop methods for using their calculators to obtain sums, differences, products, and quotients that exceed their calculators' ordinary capacities to handle large numbers.

Chapter 4: Extending Number and Operations Activities

Such a project has great potential to extend students' understanding of number and operations.

The activity takes students in a different direction, however. It asks them to determine patterns in the last digits of large products. In the process, the students discover, for instance, that the last digit of a perfect square is never 2, 3, 7, or 8. It is important for students to explain, as the activity asks them to do, why this and other discoveries are true.

In general, when students examine the first steps in the standard multiplication algorithm applied to a number like $(23{,}456{,}789)^2$, they quickly see that the last digit in the product is 1, or the last digit in 9×9. They should understand this result as an instance of a general property: the last digit in a product of two integers is the last digit in the product of their last digits. For example, $43 \times 38 = (40 + 3)(30 + 8)$ $= 1200 + 90 + 320 + 24 = 1610 + 24 = 1630 + 4$, and the last digit, 4, is the last digit in the product 3×8.

Part 2 concludes by extending the investigation of last digits to perfect cubes and other powers. Students discover patterns that allow them to complete such tasks as finding the last digit in 2007^{2007}, a number beyond the capacity of any calculator to evaluate.

Part 3—"The Singles Club"

Studying the last digits in perfect squares and other powers of integers leads smoothly to an investigation of the number system of integers modulo 10, or "clock arithmetic modulo 10." Like the activity Rock Around the Clock in chapter 3, part 3 of Perfect Squares takes a comparatively simple, fairly brief look at modular arithmetic as it applies to a particular case, in contrast to classical approaches, which begin with discussions of congruence modulo n and equivalence classes.

In this part of the activity, the students focus exclusively on the single-digit integers from 0 to 9. The activity sheet explains that these numbers compose the membership of the Singles Club, an exclusive club that exists in Math Land and is closed to outsiders under the club's special rules for addition and multiplication. These rules define two new operations: "last-digit addition" and "last-digit multiplication." The Singles Club has registered $\oplus$ and $\otimes$ as exclusive symbols for these operations. To perform them, one simply adds or multiplies in the usual fashion and then selects the last digit in the result.

The set of integers from 0 to 9 is closed under last-digit addition and last-digit multiplication. The students learn, moreover, that these operations are commutative and associative with the usual identity elements. They also find that last-digit multiplication distributes over last-digit addition. In the language of abstract algebra, the number system is a *commutative ring with unity*.

The students also discover many differences between the last-digit system and the integers, including the following:

- Every element in the last-digit system is a perfect cube.
- Most numbers in the last-digit system are zero divisors (in other words, the last-digit system includes nonzero integers whose product is 0; for example, $2 \times 5 = 0$).
- No prime numbers exist in the last-digit system (that is, every number in the system has at least one other divisor in addition to itself and 1).

In Math Land, closed means that the sum of any two singles in the Singles Club has to be a member of the club. The product of any two singles also has to be a member of the club.

- The perfect squares in the last-digit system are 0, 1, 4, 5, 6, and 9 (students should note that these numbers are the last digits in the perfect squares in the integers).

Assessment

In most instances, students can find more than one way to explain and solve the problems that the activity poses. Look for the multiple paths that your students take in their work. Moreover, you can increase their awareness of the possibilities if you have them work in groups of two or three and give each group an opportunity to share its results with the entire class.

You will probably find that your students use informal means more often than algebra to explain their results. For example, some students might make an argument like the following for the fact that the exponents are even in the prime factorizations of perfect squares: "When I factor a number into the product of two numbers, I'm separating its prime factors into two sets. But I know that the prime factors of a perfect square must have a partition in which both sets of prime factors are identical, so I know for sure that I'll have an even number of prime factors of each type."

Encourage these informal explanations, relating them whenever you can to more formal algebraic explanations that your students can use instead. For example, the informal explanation in the preceding paragraph is closely related to an argument from the laws of exponents. The solutions provide algebraic arguments that your students should be able to understand.

Like the earlier activity Number Triangles, the activity Perfect Squares lends itself to many extensions that offer opportunities for you to assess your students' understanding. For example, you can have your students look at other classes of polygonal numbers. In part 1 of Perfect Squares, your students should have discovered that for squares, $S_n - S_{n-1} = 2n - 1$, the nth odd number. The other polygonal numbers have equally interesting recursive relationships that the students can explore. In addition to investigating interesting properties that the numbers have in their own right, your students can discover the close ties that some of these numbers have to the perfect squares.

For example, the sequence of triangular numbers starts with 1, 3, 6, 10, 15 (see fig. 4.5). By adding consecutive pairs of triangular numbers, your students will generate the sequence 4, 9, 16, and 25, which they will recognize as the sequence of square numbers starting at 4. In general, if T_n is the nth triangular number and S_n is the nth square number, then $S_n = T_n + T_{n-1}$.

Students can take this result further. If P_n is the nth pentagonal number, then $P_n = S_n + T_{n-1} = T_n + 2(T_{n-1})$. If H_n is the nth hexagonal number, then $H_n = P_n + T_{n-1} = S_n + 2(T_{n-1}) = T_n + 3(T_{n-1})$. Students will find that the pattern continues for higher polygonal numbers.

Where to Go Next in Instruction

Like Number Triangles, Perfect Squares illustrates the potential of a deceptively simple but actually very rich context in number and operations. A curriculum that emphasizes the elements of the Process

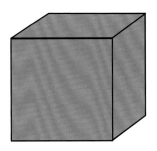

Under the operation ⊗, each member of the Singles Club is a perfect cube:

$0 = 0^3$
$1 = 1^3$
$2 = 8^3$
$3 = 7^3$
$4 = 4^3$
$5 = 5^3$
$6 = 6^3$
$7 = 3^3$
$8 = 2^3$
$9 = 9^3$

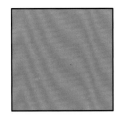

Under the operation ⊗, six members of the Singles Club are perfect squares:

$0 = 0^2$
$1 = \{1^2, 9^2\}$
$4 = \{2^2, 8^2\}$
$5 = 5^2$
$6 = \{4^2, 6^2\}$
$9 = \{3^2, 7^2\}$

Chapter 4: Extending Number and Operations Activities

Fig. **4.5.**
Geometric representations of the first five triangular, square, pentagonal, and hexagonal numbers

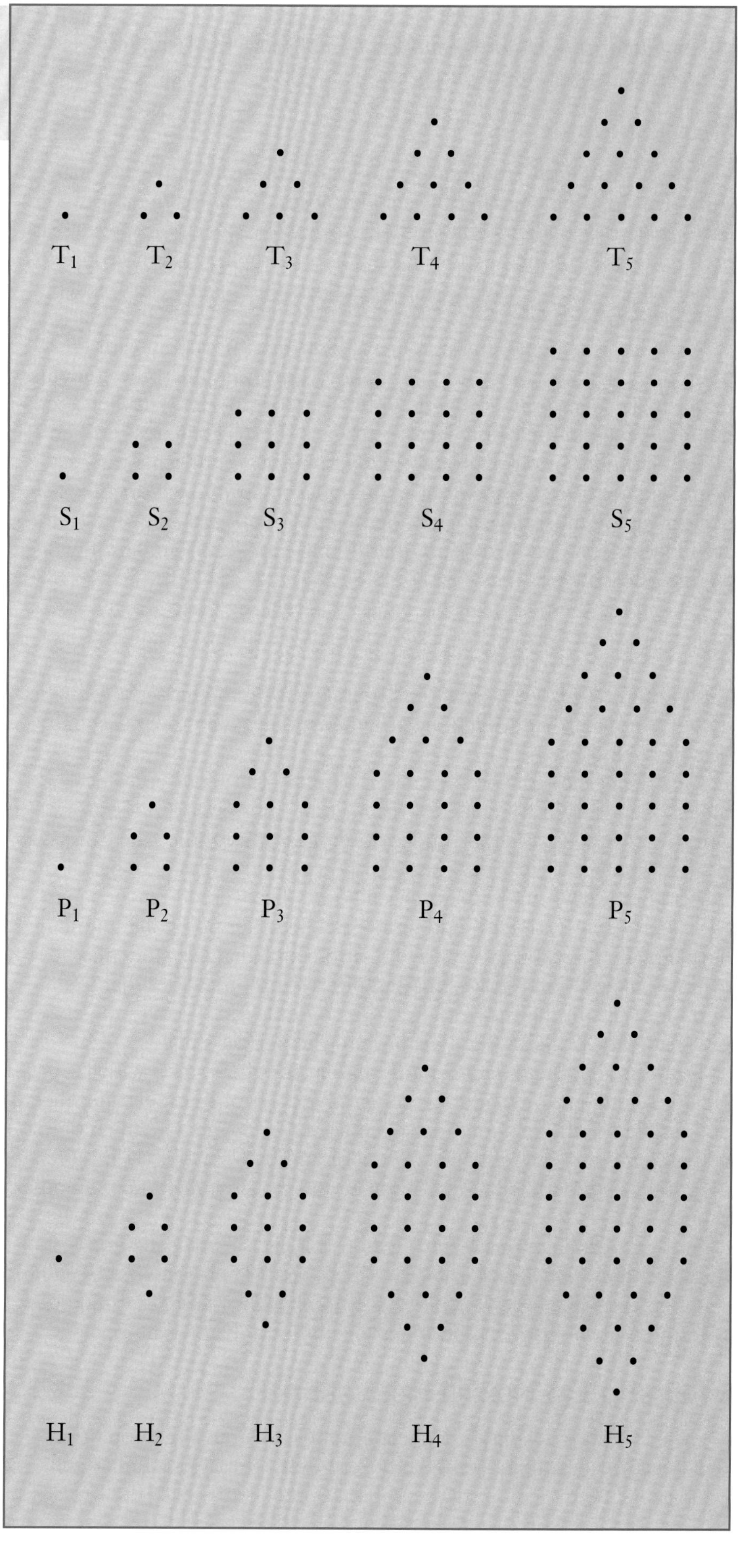

T_1 T_2 T_3 T_4 T_5

S_1 S_2 S_3 S_4 S_5

P_1 P_2 P_3 P_4 P_5

H_1 H_2 H_3 H_4 H_5

Standards encourages students to progress steadily from the basic mathematics that a context offers to deeper, more far-reaching, and more richly rewarding learning opportunities that lie beyond it. The next activity, Flooding a Water World, offers another such context.

Flooding a Water World

Goal

- Explore a counting problem that leads to several powerful results in graph theory and geometry

Materials and Equipment

For each student—

- A copy of the activity sheet "Keeping It Legal"
- A copy of the activity sheet "Other Realms, Other Regions"
- (Optional) Two Styrofoam balls (4–6 inches in diameter), 10–15 rubber bands, and 10–15 pushpins

pp. 141–45; 146–48

Discussion

This counting investigation has two parts that take the students step by step from a relatively simple and concrete consideration of a hypothetical world to the development of some important notions from the realms of geometry and graph theory. In part 1, the students explore a world that is covered by water except where the world's managers create dry, habitable regions, called *cantons*, by constructing connected networks of dikes and towers. In part 2, the students apply their conclusions to the more purely geometrical tasks of determining the relationships among the numbers of faces, vertices, and edges in a complex polyhedron and counting the number of regions in a circle divided by line segments connecting n points on the circle.

Part 1—"Keeping It Legal"

Part 1 introduces the water world and enumerates the laws that govern the construction of dikes in two kinds of network—a network without cantons and a network with cantons. The students acquaint themselves with the laws of the water world and verify that two sample networks—one with cantons and one without cantons—are legal (see figure 4.6).

A brief exploration shows the students that in a network without cantons, the number of dikes always exceeds the number of towers by one. They can easily prove this result by noting that such a network can begin with just one tower to which dikes are added. Because in this network no dikes enclose cantons, each new dike always has an end attached to a new tower, which is not attached to any other dike. Thus, in a network with no cantons, the number of towers is always one more than the number of dikes.

This relationship becomes important later in the investigation. Meanwhile, the students discover no corresponding relationship between the number of dikes and the number of towers in a network with cantons. But they quickly become aware of another relationship, which also becomes significant later. The water world defines a *dike load*

a

b

Fig. 4.6.

Two legal water-world networks, (a) with cantons and (b) with no cantons, shown on models made with Styrofoam balls, rubber bands, and pushpins

---

The idea for the water world in this activity comes from *Transformational Geometry: An Introduction to Symmetry* by George E. Martin (1982).

as the number of dikes that intersect at a tower, and the managers assert that the sum of a network's dike loads is always twice its number of dikes.

This claim is also easy for the students to verify. Every dike connects two different towers and so gets counted in two dike loads. Thus, every dike appears twice in the sum of the dike loads. This fact corresponds to a theorem in graph theory. According to this theorem, the sum of the degrees of the vertices in a *multigraph*—a graph in which more than one edge can connect two vertices—is twice the number of edges. The students use this idea in part 2.

Part 1 continues by probing what happens if the managers of the water world decide to flood all the cantons in a legal network. The water world has laws that govern the removal of dikes as well as their construction. How many dikes must the managers remove to flood all the cantons while preserving a legal—that is, a fully connected—network? How many dikes will remain?

By experimenting with networks of their own devising, comparing their work, and discussing their results, the students quickly discover that for any network with cantons,

Number of removed dikes = Number of cantons.

They also realize that to count the number of remaining dikes, they can use the relationship that they identified earlier between the number of dikes and the number of towers in a network without cantons:

Number of towers = Number of dikes + 1.

This relationship is relevant since flooding all the cantons in a network transforms it into a network without cantons. Thus,

Number of remaining dikes = Number of towers – 1.

The students should note that removing a dike from the perimeter of a canton does not disconnect either of the towers that previously served as the dike's terminal points. A canton comes into being when a system of dikes and towers encloses a region. In such a system, any two towers on the boundary of a canton are always attainable by two disjoint paths. Because a detour, or alternate route, always exists for any path that previously traversed a now-removed dike, all the towers in the network remain connected. This result also corresponds to an important result in graph theory. If an edge in a connected multigraph is removed from a cycle, then the graph stays connected.

The students' work has now prepared them to discover the main result in part 1—the idea that the number of dikes in any water-world network is one less than the total number of cantons and towers in the network. The students can use basic algebra to arrive at this result. The number of dikes in any network is equal to the sum of the dikes removed and the dikes remaining. Thus,

Number of dikes = Number of removed dikes +
Number of remaining dikes

= Number of cantons +
Number of towers – 1.

The following laws govern the construction of dikes in the water world:

- All dikes must begin at one tower and end at a different one.
- No dike may join more than two towers.
- Two dikes may intersect only at a tower.
- More than one dike may connect two particular towers.
- All dikes must be connected in a network so that citizens may travel on top of them from any tower to any other tower in the network.

◆ ◆ ◆

The following laws govern the removal of dikes from a network with cantons:

- A dike may be removed only if doing so will lead immediately to the flooding of a dry region.
- A dike that has water on both sides may not be removed, since such an action could inadvertently disconnect a network.

A *graph* is a set of *vertices and edges* with each edge containing exactly two distinct vertices. If two vertices can be contained by more than one edge, the graph is a *multigraph*. The following diagrams represent vertices as dots and edges as curved lines.

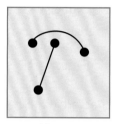

(a) Graph

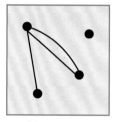

(b) Multigraph

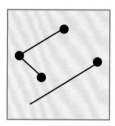

(c) Nongraph

When each pair of consecutive vertices in a sequence of vertices $v_0, v_1, v_2, \ldots, v_k$ is connected by an edge, the sequence of vertices and the designated edges form a *path* from v_0 to v_k. A path from a vertex to itself, with at least two edges and no edge occurring more than once in the path, is a *cycle*.

This is really another result from graph theory. In a connected planar graph, the number of edges is one less than the number of enclosed regions plus the number of vertices.

Part 2—"Other Realms, Other Regions"

Part 2 turns to the realm of pure geometry for two counting problems that the students can solve by applying ideas from the water world in part 1. The first problem calls for a formula that relates the numbers of faces, edges, and vertices in any convex polyhedron. The second problem asks for a count of the regions in a circle cut by line segments when no more than two line segments intersect at any point in the interior of the circle.

The first problem becomes very approachable when the students think of a complex polyhedron, such as a cube, as a water world whose towers and dikes, correspond, respectively, to the vertices and edges of the polyhedron. To complete the analogy, the students must designate one face of the polyhedron to represent water and let the rest represent cantons created by a network of dikes and towers. By transferring the concept of the water world to this new context, the students can easily apply findings from their water-world investigation to determine a formula that relates the numbers of faces, edges, and vertices in any convex polyhedron.

To derive the formula, the students work directly from the formula that they developed in the last step of part 1:

Number of dikes = Number of cantons + Number of towers – 1.

In translating this formula from the water world of dikes (D), towers (T), and cantons (C) to the purely geometric world of edges (E), vertices (V), and faces (F), the students must recognize that D equals E, and T equals V, but C is equal to $F - 1$, since they designated one face of the polyhedron as water in simulating the water world. By substitution, their formula from part 1, $D = C + T - 1$, thus becomes $E = F - 1 + V - 1$, or $E = F + V - 2$. This is in fact Euler's formula for the number of faces, edges, and vertices in a convex polyhedron.

To validate Euler's formula experimentally, the students might collect containers in as many different shapes as they can and bring them to class. Packages in grocery and department stores offer a surprising number of polyhedra that the students can find and experiment with in the classroom.

In the second problem, the students consider the situation when someone chooses n points on a circle and connects each pair of these points by the chord that the pair determines, thus dividing the circle into regions. The problem includes a restriction—the points on the circle must be situated so that at most two of the resulting chords intersect at any point in the interior of the circle. For example, if six points are placed on the circle so that they form the vertices of a regular hexagon, three of the chords will intersect at the center of the circle, as the three dotted chords in figure 4.7 illustrate. This configuration of points violates the conditions of the problem.

The activity sheet suggests a way for the students to model the problem as a question about the number of cantons in a water-world network. They can think of the points where chords intersect as the towers of the network, and the line segments and arcs that connect the intersection points as dikes in a legal network. The problem then reduces to finding the number of cantons in the network, since each region of the circle will correspond to a canton.

The activity sheet also reminds the students of the important relation

Number of cantons = Number of dikes – Number of towers + 1.

To find the number of towers, the students verify that every choice of four points on the circle determines exactly one of the points where the chords intersect on the interior of the circle (see fig. 4.8; the activity page provides this information as a hint). Therefore, if n points are placed on the circle according to the rule that no three chords intersect at the same point, then there will be $\binom{n}{4}$ intersection points on the interior of the circle. Together with the n points on the circle (which are also intersection points of the chords), there are $\binom{n}{4} + n$ points where chords intersect. Hence, the number of towers in the water-world network that corresponds to the circle and its chords is $\binom{n}{4} + n$.

To find the number of line segments and arcs that correspond to dikes, the students can readily determine that each of the n points on the circle is the endpoint of $n - 1$ line segments and 2 arcs that represent dikes. Furthermore, each intersection point inside the circle is the endpoint of four of the line segments that represent dikes. Since the sum of the dike loads in any legal water-world network is twice the number of dikes, the students can deduce the following formula for the number of dikes in the water-world network corresponding the circle and its chords:

$$\text{Sum of the dike loads} = n(n+1) + 4 \times \binom{n}{4};$$

therefore,

$$\text{Number of dikes} = \frac{n(n+1) + 4 \times \binom{n}{4}}{2}.$$

By substituting the expressions that they have found for the number of dikes and the number of towers into the formula

Number of cantons = Number of dikes – Number of towers + 1,

the students can arrive at the formula for the number of regions (*R*) in the circle:

$$R = \binom{n}{4} + \frac{n^2}{2} - \frac{n}{2} + 1 = \frac{n(n-1)(n-2)(n-3)}{4 \times 3 \times 2 \times 1} + \frac{n^2}{2} - \frac{n}{2} + 1$$

$$= \frac{n^4}{24} - \frac{n^3}{4} + \frac{23n^2}{24} - \frac{3n}{4} + 1.$$

(See Conway and Guy [1996] for a nice explanation of this classic puzzle.)

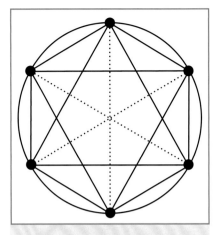

Fig. **4.7.**

Six points on a circle at the vertices of a regular hexagon, with all the chords that the six points determine

The problem includes a restriction—the points on the circle must be situated so that at most two of the resulting chords intersect at any point in the interior of the circle.

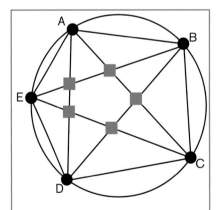

Number of points on the interior of the circle where chords intersect = 5, or $\binom{5}{4}$.

Number of disjoint regions of the circle determined by the chords = 16

Fig. **4.8.**

Each selection of four points on the circle (for example, *E*, *B*, *C*, and *D*) determines a cyclic quadrilateral (in the example,) whose diagonals determine one intersection point (in the example, *M*) inside the circle.

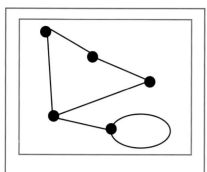
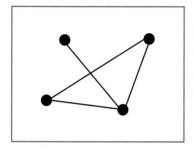
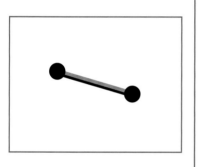

Fig. **4.9.**

Examples of configurations that do not qualify as legal water-world networks

Parker (2205; available on the CD-ROM) discusses the use of inductive reasoning to generalize about the results of partitioning a circular region with chords.

Assessment

In part 1, "Keeping It Legal," students need to be able to distinguish between legal and illegal networks. The activity provides opportunities for learning the distinction, but you might ask your students to pause after completing step 1 to consider several cases that do not satisfy the laws of the water world. This break will give you a chance to assess their understanding of this important distinction. Figure 4.9 includes examples that do not qualify as legal networks because (a) one of the dikes connects a tower to itself, (b) one of the dikes intersects another dike at a point where there is no tower, and (c) two dikes run side by side between the same two towers with no space in between. (If at any point there is no space between two dikes, then they must intersect at that point, and hence there has to be a tower there.)

"Keeping It Legal" also helps students discover several general relationships among the numbers of dikes, canton, and towers in any legal network. These relationships are not hard to discern. However, your students might have more trouble explaining why the relationships hold for all legal networks. Frequently, students see a pattern and know intuitively that it is true in all cases, yet pinpointing or expressing any reason for it is difficult.

For example, in a legal network with no cantons, the number of towers must be one more than the number of dikes. How could students explain that this is true? The explanation is essentially a proof by induction. Even though you should not expect your students to discover such a proof on their own, they should grasp the essential idea of the proof—that is, that adding a dike to an existing network in such a way that no canton is created requires adding a tower, as well.

A good way to assess your students' understanding of the essential ideas of the activity Flooding a Water World is to monitor their responses to problem 1 of part 2, "Other Realms, Other Regions." This problem basically calls for a transfer of the ideas that the students developed in "Keeping It Legal" to a similar but more abstract context.

Conclusion

When students work in number contexts like the ones in this chapter's activities, proof becomes a natural, integral part of the problem-solving process. This is surely one of the greatest benefits of offering such problems to students. The deceptively simple contexts of these activities lead seamlessly to demonstrations of significant mathematical results.

Working together in such contexts, students learn how thoroughly intertwined the elements of the mathematical process actually are. This insight accompanies the new mathematical knowledge that they build. They discover that solving problems, communicating their thinking, making connections, representing their ideas clearly, and—last but not least—sharing reasons for and proofs of their evolving ideas are essential to making new discoveries about number and operations.

Navigating through Number and Operations

GRADES 9–12

Looking Back and Looking Ahead

Problems involving whole numbers and their operations may offer students the most appropriate contexts in which to develop a deep understanding of number systems and to learn how to prove basic results.

The activities and problems for grades 9–12 in this book have illustrated central themes that *Principles and Standards for School Mathematics* (NCTM 2000) outlines in the Number and Operations Standard. The most important advance that students make in high school is the development of an understanding of the real number system. This understanding lays a foundation for much of the work that students are likely to do in mathematics at the high school and college levels.

Students' grasp of the real numbers should have roots in a firm understanding of rational and irrational numbers and should connect with fundamental notions in the other content strands elaborated in *Principles and Standards*: algebra, geometry, measurement, and data analysis and probability. The activities in this book highlight some of these connections, including links to geometry, the theory of equations in algebra, and real-world applications of measurement.

Principles and Standards also calls for students to engage in reasoning and proof in the context of number systems. Although it is sufficient for middle school students to be fluent in the use of rational numbers and their operations, high school students must be able to reason and prove simple results about the properties of rational numbers—especially whole numbers. In fact, problems involving whole numbers and their operations may offer students the most appropriate contexts in which to develop a deep understanding of number systems and to learn how to prove basic results. Several activities in this book illustrate the potential

To appreciate the steadily unfolding, increasingly complex understanding of number and operations that students develop, see the activities in Navigating through Number and Operations in Prekindergarten–Grade 2 *(Cavanaugh et al. 2004),* Navigating through Number and Operations in Grades 3–5 *(Duncan et al. 2006), and* Navigating through Number and Operations in Grades 6–8 *(Rachlin et al 2006).*

The CD-ROM includes a sampling of articles from the journal *Mathematics Teacher* to illustrate some of the possibilities. See "From Exploration to Generalization: An Introduction to Necessary and Sufficient Conditions" (Bonsangue and Gannon 2003), "An Odd Sum" (Shiflett and Shultz 2002), "Recursion and the Central Polygonal Numbers" (Miller 1991), and "Star Numbers and Constellations" (Francis, 1993).

of whole numbers to supply opportunities for reasoning and proof within the study of number and operations.

To develop an understanding of the real numbers and to begin proving simple results about the rational numbers, students in grades 9–12 need curricula and instruction that build on the ideas that they have mastered in middle and elementary school. This book cannot possibly communicate the scope of the steady, interconnected development that *Principles and Standards* envisions for number and operations from prekindergarten through grade 12. However, the introduction offers an overview of students' progress, outlining the principal skills and knowledge that students acquire over the years. This introduction can help to put the number and operation goals for grades 9–12 into perspective.

For additional details about the growth of understanding and the connections from one level to the next, readers may want to delve into the activities that the Navigations Series presents for number and operations in prekindergarten–grade 2, grades 3–5, and grades 6–8. These activities illustrate the ways in which students' understanding grows and deepens from stage to stage.

Numerous articles in the journal *Mathematics Teacher* also support the ideas about number and operations in this book. More important, articles in *Mathematics Teacher* often go farther than the activities here. Several articles that appear on the CD-ROM illustrate some of the possibilities. For example, Bonsangue and Gannon (2003) show how working with whole numbers and their operations can develop students' reasoning abilities. Shiflett and Shultz (2002) and Miller (1991) demonstrate how to apply such reasoning at deeper levels by using algebra to prove interesting results about whole numbers and figurate numbers. Francis (1993) gives a glimpse of the array of number-theory results that curious high school students can explore.

A number of important curricular issues loom ahead for number and operations. Educators in the twenty-first century must keep abreast of developments in technology and their effects on the mathematics curriculum in grades 9–12. While the technology continues to evolve with breathtaking speed, one fact remains clear and unchanged: the mathematics curriculum must also evolve as technology transforms our society and its needs. However, another fact seems equally clear: whatever shape the curriculum takes, and whatever applications and mathematical understandings emerge as important goals for our students, number and operations will remain the core of the prekindergarten—grade 12 mathematics curriculum.

Navigating through Number and Operations

Navigations Series

Grades 9–12

Appendix

Blackline Masters and Solutions

When Fractions Are Whole

Name _____

Consider the positive rational numbers. These numbers can be represented by fractions $\frac{p}{q}$, where p and q are positive integers and $q \neq 0$. Mathematicians sometimes call such fractions *rational fractions*. A rational fraction $\frac{p}{q}$ is in simplest form when the greatest common divisor of p and q is 1. The square of a fraction $\frac{p}{q}$ is

$$\frac{p}{q} \times \frac{p}{q}, \text{ or } \frac{p^2}{q^2}.$$

The following list identifies some of the possible properties of a rational fraction:

1.	The fraction is in simplest form.
2.	The fraction is not in simplest form.
3.	The fraction is in simplest form with denominator = 1.
4.	The fraction is in simplest form with denominator ≠ 1.
5.	The fraction is equivalent to a positive integer.
6.	The fraction is not equivalent to a positive integer.
7.	The fraction's square is equivalent to a positive integer.
8.	The fraction's square is not equivalent to a positive integer.
9.	The fraction's square is in simplest form.
10.	The fraction's square is not in simplest form.
11.	The fraction's denominator is a divisor of its numerator.
12.	The fraction's numerator is a divisor of its denominator.

You can investigate rational fractions by searching for examples that exhibit particular combinations of these properties. Refer to the list to solve the problems in steps 1–3. Note that examples may not exist for all combinations!

1. For each case, give an example of a rational fraction with the specified characteristics.

   a. The fraction satisfies properties 1 and 6.

   b. The fraction satisfies properties 1 and 5.

Navigating through Number and Operations in Grades 9–12

When Fractions Are Whole (continued)

Name _____

    c. The fraction satisfies properties 2 and 5.

    d. The fraction satisfies properties 2 and 7.

    e. The fraction satisfies properties 1, 11, and 12.

2. For each case, give an example of a rational fraction with the specified characteristics. If this is not possible, explain why no such fraction exists.

    a. The fraction satisfies properties 5 and 10.

    b. The fraction satisfies properties 3 and 6.

    c. The fraction satisfies properties 1 and 10.

When Fractions Are Whole (continued)

Name _____

d. The fraction satisfies properties 4 and 5.

e. The fraction satisfies properties 6 and 11.

f. The fraction satisfies properties 2 and 9.

g. The fraction satisfies properties 3 and 8.

3. Referring again to the list of properties of a rational fraction, complete the following statements to make them true. Compare your completed statements with those of your classmates.

    a. **Fact 1:** If a fraction is in simplest form, then the square of the fraction _____

    _____

    b. **Fact 2:** If a fraction is in simplest form and is equivalent to a positive integer, then its denominator _____ .

4. Suppose that N is a positive integer and $\sqrt{N} = \frac{p}{q}$, a rational fraction in simplest form. Use facts 1 and 2 from step 3 to help you answer the following questions:

    a. Since $\sqrt{N}$ is equal to $\frac{p}{q}$, what does $\left(\frac{p}{q}\right)^2$, or $\frac{p^2}{q^2}$, equal?

When Fractions Are Whole (continued)

Name _____

   b. Is $\dfrac{p^2}{q^2}$ in simplest form? _____ Why, or why not?

   c. What can you conclude about the denominator in $\dfrac{p^2}{q^2}$? *Hint:* Remember to refer to your facts from step 3.

   d. What can you conclude must be true about *N*?

   What positive integer values can *N* have?

5. Using your work from step 4, complete the following sentence to produce a true statement:

   When *N* is a positive integer, if $\sqrt{N} = \dfrac{p}{q}$, a rational fraction in simplest form, then *N* _____ .

6. What can you conclude about the square roots of positive integers that are not perfect squares?

7. Explain why $\sqrt{2}$ is irrational.

Designing a Line

Name _____

In 2000 B.C., the ancient Egyptians used a system based on a *cubit* and a *palm* to measure length. A cubit was the distance from the elbow down the forearm to the tip of the middle finger. A palm was the width of the hand at the base of the four fingers ($\approx$ 3 inches). One of the world's oldest mathematical documents, the Rhind Papyrus, treats a palm as one-seventh of a cubit.

Suppose that a modern-day archaeological association, the Ancient Egyptian Revival Society, wants to recreate this system for an exhibit and has asked you to develop a number line whose basic unit is the palm. For the task, you have a strip of ribbon, fabric tape, or paper, as well as a 3-by-5-inch index card to approximate a palm. The Ancient Egyptian Revival Society demands authenticity in every aspect of its recreations. In keeping with ancient methods, you must not use a calculator or a ruler in your task!

1. *a.* Mark a point near one end of your strip, and label it 0.

   b. Positioning the end with 0 on your left, mark a point one palm to the right of 0, and label it 1.

   c. On your number line, mark the positions, as on a ruler, of the following lengths in palms:

   $$2, 3, 4, 5, 6, 7, \frac{1}{6}, \frac{2}{6}, \frac{3}{6}, \cdots, \frac{41}{6}, \sqrt{2}, \sqrt{3}, \sqrt{5}, \pi, 2\pi, \frac{\pi}{2}, \frac{3\pi}{2}.$$

   Note: Remember that you cannot use a calculator or a ruler. Use geometry instead! Your teacher can give you an additional 3-by-5-inch card, and possibly some construction paper or tagboard, if you need it as you work.

2. Now that you have represented real numbers of palm-lengths, both rational and irrational, as points on your line, explain how to find the points that correspond to the following arithmetic results in palms:

   a. $\dfrac{1+\sqrt{5}}{2}$ b. $\pi - \sqrt{2}$ c. $\pi \times \sqrt{2}$ d. $\dfrac{\pi}{\sqrt{2}}$

3. Work with a partner in your group and use his or her number line along with your own line to form a pair of perpendicular lines, as in figures 1 and 2.

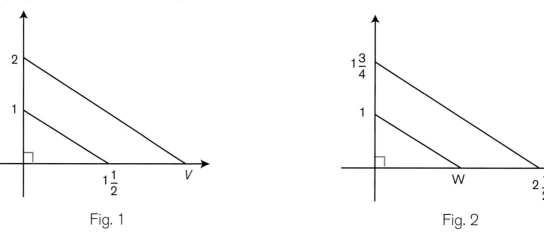

Fig. 1   Fig. 2

Designing a Line (continued)

Name _____

Assume that the line segments forming the hypotenuses of the right triangles in the diagrams are parallel. What are the values of the points labeled *V* and *W* on the horizontal number lines?

V = _____ palms   W = _____ palms

4. *a.* Did your results in step 3 depend on the fact that you and your partner joined the two number lines to form a right angle?

   b. Would other angles have worked just as well and have given you the same results? _____
   Why, or why not?

5. Examine figures 3 and 4.

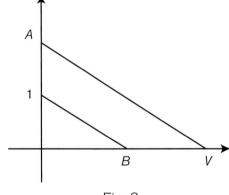

Fig. 3

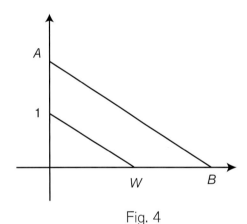

Fig. 4

For any value $A \neq 0$ on your vertical number line and any value $B \neq 0$ on your horizontal number line, assuming that the line segments forming the hypotenuses of the right triangles are parallel, how could you use *A* and *B* to find the values of the points *V* and *W* on the horizontal number lines?

V = _____ palms   W = _____ palms

6. Apply the methods that you developed in steps 3–5.

   a. Show that $\sqrt{2} \times \pi = \pi \times \sqrt{2}$ palms.

Designing a Line (continued)

Name _____

    b. What does your work in 6(*a*) suggest about the multiplication of points on the number line in general? Justify your answer.

    c. Show that 1 is the multiplicative identity. That is, show $1 \times A = A \times 1 = A$.

    d. Show that every point $A \neq 0$ has a multiplicative inverse by finding $\frac{1}{A}$ on your number line.

7. If you think of real numbers as points on a line, then you can carry out and interpret arithmetic operations on real numbers geometrically. But what if you think of real numbers as infinite decimals? How would you perform the arithmetic operations then? Discuss with your group how would you perform operations like those in step 2. For example, how would you multiply $\pi \times \sqrt{2}$; that is, how would you multiply 3.141592653589… × 1.4142135623…?

8. *a.* Are there other real numbers that you would choose to show in palms on your number line for the exhibit of the Ancient Egyptian Revival Society?

    b. Are there any numbers whose inclusion might be unrealistic on a number line that recreates a measuring system from 2000 B.C.?

Trigonometric Target Practice

Name _____

The graph shows a line *l* that passes through the origin and through point (*h*, *k*) and makes an angle of θ radians with the positive *x*-axis.

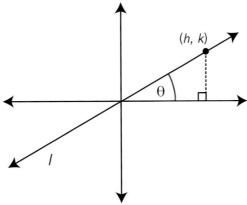

1. Consider the point (*h*, *k*) and the angle θ.

   a. In terms of *h* and *k*, what is tan θ?

   b. What is the equation of the line?

   c. What is the relation between the equation of the line and tan θ?

2. A *lattice point* is a point that falls at the intersection of grid lines on a Cartesian grid. Thus, its coordinates are integers. Assume that line *l* passes through the origin.

   a. For line *l* to pass through a second lattice point (*p*, *q*), where *p* ≠ 0, what must be true about the slope of *l*? (Explain your answer.)

   b. If line *l* passes through the origin as well as through a lattice point (*p*, *q*), where *p* ≠ 0, what other lattice points does it pass through? (Explain by using an example.)

Navigating through Number and Operations in Grade 9–12

Trigonometric Target Practice (continued)

Name _____

3. Use a graphing calculator with the grid turned on and with angles measured in radians.

   a. Graph the line $y = \tan\left(\dfrac{5}{4}\right) \cdot x$. Does the line appear to pass through any lattice points other than (0, 0)? _____ Describe your findings.

   b. Try to "hit" a second lattice point in addition to (0, 0) by replacing $\dfrac{5}{4}$ in the equation for the line in 3(a) with other nonzero rational fractions.

   Record the equation of each new line that you graph, and tell whether the line appears to pass through any lattice points besides (0, 0).

   c. When the line $y = \tan(\theta) \cdot x$ is extremely close to a lattice point (p, q), what is a good approximation for the value of $\tan \theta$?

4. From your attempts to "hit" a lattice point in step 3, what do you conclude about the values of $\tan \theta$ when θ is a nonzero rational number of radians?

Counting Primes

Name _____

1. An integer is a *multiple* of 5 if and only if it can be written in the form $5 \times n$, where *n* is an integer.

   a. On the number line, label the multiples of 5 in the factored form, like 5×1.

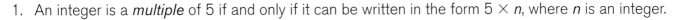

   b. Select two multiples of 5 on the number line and find their difference. Is the difference a multiple of 5?

   c. If the difference between two numbers on the number line is 1, can the two numbers both be multiples of 5? _____ Explain.

   d. Prove that no matter what two multiples of 5 you select, their difference is a multiple of 5.

2. Select a different single-digit integer greater than 1.

   a. Label its multiples on the number line as you did in step 1.

   b. Select two multiples of your new number and find their difference. Is the difference a multiple of your number?

   c. If the difference between two numbers on the number line is 1, can the two numbers both be multiples of your new number? _____ Explain.

Navigating through Number and Operations in Grade 9–12   91

Counting Primes (continued)

Name _____

    d. Prove that no matter what two multiples of your number you select, their difference is a multiple of your number.

    e. Compare your results with those of other members of your class, and make generalizations that summarize the class results.

3. If M and $M + 1$ are two consecutive integers on the number line, both are multiples of 1. In other words, 1 is a divisor of both M and $M + 1$. Are there any other positive integers that can be divisors of both M and $M + 1$? _____ Justify your answer.

4. An integer $p > 1$ is *prime* if it has exactly two positive divisors. For example, 7 is a prime number since its only positive divisors are 7 and 1. An integer n that is greater than 1 is *composite* if it has more than two positive divisors. For instance, 6 is a composite number since it has four positive divisors: 1, 2, 3, and 6.

    a. On the number line above, label all the integers and circle all the prime numbers that you can.

    b. Make sure that you have circled the same numbers as your classmates.

5. Randomly choose a set of up to six prime numbers from the primes that you circled in 4(*a*).

    a. List the elements of your set.

Counting Primes (continued)

Name _____

b. Multiply your selected prime numbers together and add 1. Call your resulting number N.

$N =$ _____

c. Use a calculator or other technology to find the prime factorization of N. (The illustration shows how the screen will look on some calculators. For the prime numbers 2, 5, 7, 19, and 23, for example, $N = 2 \times 5 \times 7 \times 19 \times 23 + 1$, or 30591. The calculator's factor command produced the prime factorization of $N = 30591$: $3 \times 3 \times 3 \times 11 \times 103$.)

Record the prime factorization of your number N.

6. Answer the following questions on the basis of your work in step 5.

   a. Is your number N a prime number? _____

   b. What prime numbers are factors of N? _____ Do any of them match numbers in the set of prime numbers that you used to form N?

   c. Divide N by any one of the prime numbers that you selected. (Divide by hand, or use another method that doesn't use a calculator or computer.) What remainder do you obtain? _____

   d. Compare your results with your classmates' results, looking for patterns and recording any generalizations that you think are true.

Counting Primes (continued)

Name _____

7. Suppose that $S = \{p_1, p_2, p_3, p_4, p_5, p_6\}$ is a set of six prime numbers.

   a. Why do the two integers $p_1 \times p_2 \times p_3 \times p_4 \times p_5 \times p_6$ and $p_1 \times p_2 \times p_3 \times p_4 \times p_5 \times p_6 + 1$ have no common prime factors?

   b. Can you generalize your answer in 7(a) to handle sets of prime numbers of arbitrary size? _____ Explain your generalization.

   c. If $S = \{p_1, p_2, p_3, \ldots, p_n\}$ is any finite set of prime numbers, what method could you use to find one or more prime numbers that are not in set S?

8. Does the set of all prime numbers have a finite number of elements, or must it have an infinite number of elements? _____ Justify your answer.

Adding Complex Numbers

Name _____

Complex Numbers and Matrices—Part 1

1. Find the following sums of complex numbers:

   a. $(3 + 4i) + (6 + 3i) =$ _____

   b. $(2 - 4i) + (-3 + i) =$ _____

   c. $(-5 - 2i) + (4 - 3i) =$ _____

2. You can represent the sum of $(3 + 4i)$ and $(6 + 3i)$ from step 1(a) as the addition of the respective coordinates of the points (3, 4) and (6, 3). This addition gives a new ordered pair, $(3 + 6, 4 + 3)$. The same is true for the other sums in step 1.

   a. Complete the table to show each sum.

(x_1, y_1)	(x_2, y_2)	$(x_1 + x_2, y_1 + y_2)$
(3, 4)	(6, 3)	

   b. Use your own examples to explain how all complex numbers and their sums correspond to ordered pairs of real numbers and their sums.

3. Vectors can represent the sum of two complex numbers as the addition of ordered pairs. Again suppose that you are adding $3 + 4i$ and $6 + 3i$. The diagram shows the ordered pairs in this sum as the endpoints of two vectors, $\overrightarrow{(3,4)}$ and $\overrightarrow{(6,3)}$. The endpoint of the vector sum, or the resultant vector, $\overrightarrow{(9,7)}$, corresponds to the ordered pair that we obtain by adding the coordinates.

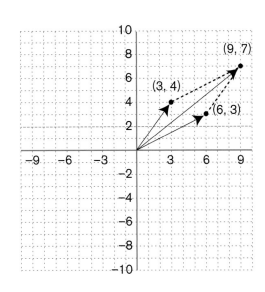

Navigating through Number and Operations in Grade 9–12 95

Adding Complex Numbers (continued)

Name _____

a. Considered together, the origin, the endpoints of the two vectors being added, and the endpoint of the resultant vector determine what kind of quadrilateral?

b. Use the other two addition problems that you represented with ordered pairs in step 2 to draw quadrilaterals that represent the vector sum of each.

$(2 - 4i) + (-3 + i)$

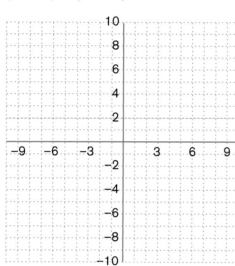

$(-5 - 2i) + (4 - 3i)$

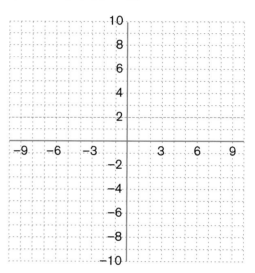

c. Use examples of your own to explain how all complex numbers and their sums correspond to vectors and their sums.

Adding Complex Numbers (continued)

Name _____

4. Add each pair of matrices and compare each sum with the sums in step 1.

   a. $\begin{bmatrix} 3 & 4 \\ -4 & 3 \end{bmatrix} + \begin{bmatrix} 6 & 3 \\ -3 & 6 \end{bmatrix} = \begin{bmatrix} & \\ & \end{bmatrix}$

   b. $\begin{bmatrix} 2 & -4 \\ 4 & 2 \end{bmatrix} + \begin{bmatrix} -3 & 1 \\ -1 & -3 \end{bmatrix} = \begin{bmatrix} & \\ & \end{bmatrix}$

   c. $\begin{bmatrix} -5 & -2 \\ 2 & -5 \end{bmatrix} + \begin{bmatrix} 4 & -3 \\ 3 & 4 \end{bmatrix} = \begin{bmatrix} & \\ & \end{bmatrix}$

5. Does adding two matrices of the form $\begin{bmatrix} a & b \\ -b & a \end{bmatrix}$ always yield a matrix of the same form? _____

   Use examples of your own to explain how all complex numbers and their sums correspond to particular 2 × 2 matrices and their sums.

Multiplying Complex Numbers

Name _____

Complex Numbers and Matrices—Part 2

1. Find the following products of complex numbers:

   a. $(3 + 4i)(6 + 3i) = $ _____

   b. $(2 - 4i)(-3 + i) = $ _____

   c. $(-5 - 2i)(4 - 3i) = $ _____

2. Multiply each pair of matrices:

   a. $\begin{bmatrix} 3 & 4 \\ -4 & 3 \end{bmatrix} \begin{bmatrix} 6 & 3 \\ -3 & 6 \end{bmatrix} = \begin{bmatrix} & \\ & \end{bmatrix}$

   b. $\begin{bmatrix} 2 & -4 \\ 4 & 2 \end{bmatrix} \begin{bmatrix} -3 & 1 \\ -1 & -3 \end{bmatrix} = \begin{bmatrix} & \\ & \end{bmatrix}$

   c. $\begin{bmatrix} -5 & -2 \\ 2 & -5 \end{bmatrix} \begin{bmatrix} 4 & -3 \\ 3 & 4 \end{bmatrix} = \begin{bmatrix} & \\ & \end{bmatrix}$

3. a. Does multiplying two matrices of the form $\begin{bmatrix} a & b \\ -b & a \end{bmatrix}$ always yield a matrix of this form? _____

   b. Use examples of your own to explain how multiplying two matrices of the form $\begin{bmatrix} a & b \\ -b & a \end{bmatrix}$, as in step 2, compares with the multiplication of complex numbers, as in step 1.

Multiplying Complex Numbers (continued)

Name _____

4. Consider the two matrices $A = \begin{bmatrix} 4 & 1 \\ 2 & -3 \end{bmatrix}$ and $B = \begin{bmatrix} -2 & 5 \\ 0 & 6 \end{bmatrix}$. Compute $A \times B$ and $B \times A$.

   Is matrix multiplication commutative in general?

5. Use the examples in step 2 to investigate multiplication with matrices of the form $\begin{bmatrix} a & b \\ -b & a \end{bmatrix}$.
   Does such multiplication appear to be commutative? _____
   Explain your conclusion.

6. The multiplicative identity for complex numbers is $(1 + 0i)$. The multiplicative identity for 2×2 matrices is $\begin{bmatrix} 1 & 0 \\ 0 & 1 \end{bmatrix}$.

   Find and compare the inverses of $(3 + 4i)$ and $\begin{bmatrix} 3 & 4 \\ -4 & 3 \end{bmatrix}$ by solving the following equations:

   a. $(3 + 4i)(a + bi) = (1 + 0i)$

   b. $\begin{bmatrix} 3 & 4 \\ -4 & 3 \end{bmatrix} \begin{bmatrix} a & b \\ -b & a \end{bmatrix} = \begin{bmatrix} 1 & 0 \\ 0 & 1 \end{bmatrix}$

   c. Compare the inverses that you obtained. Are they the same? _____ Explain.

Navigating through Number and Operations in Grade 9–12

Multiplying Complex Numbers (continued)

Name _____

7. You have explored two systems of numbers and operations with them — complex numbers of the form *a* + *bi* and matrices of the form $\begin{bmatrix} a & b \\ -b & a \end{bmatrix}$, under addition and multiplication. As a system of numbers and operations, how does the matrix system compare with the complex number system? (Make sure to identify the properties that they have in common.)

Solving Real Numbers

Name _____

Solve That Number—Part 1

You have solved equations many times. For instance, you have solved such equations as $x^2 - 6x + 8 = 0$ by finding the roots of the polynomial $x^2 - 6x + 8$. To solve the equation, you might factor $x^2 - 6x + 8$ as $(x - 2)(x - 4)$. Substituting into the original equation would give you $(x - 2)(x - 4) = 0$, and thus you would find that the solutions of the equation are 2 and 4.

This activity asks you to reverse the process, instead finding an equation that a given number solves. Think of this unusual process as "solving the number."

1. For each given number x, find an equation $P(x) = 0$, where $P(x)$ is a polynomial with integer coefficients, and x is a root of the polynomial.

   a. 5 _____

   b. −3 _____

   c. $\dfrac{2}{7}$ _____

   d. $\sqrt[3]{6}$ _____

   e. $\sqrt{12} + 1$ _____

2. Compare your method of "solving" the numbers in step 1 with the methods of other members of your class. Does one method work for all the numbers?

3. For each number x in step 1, find an equation $P(x) = 0$, where $P(x)$ is a *different* polynomial with integer coefficients, and x is a root of the polynomial.

   a. 5 _____

   b. −3 _____

   c. $\dfrac{2}{7}$ _____

Navigating through Number and Operations in Grade 9–12

Solving Real Numbers (continued)

Name _____

   d. $\sqrt[3]{6}$ _____

   e. $\sqrt{12} + 1$ _____

4. Compare your responses and the methods that you used in step 3 with those of other members of your class. Did you get the same answers? _____ Did you use the same methods?

5. Describe a method for generating many "solutions" to the numbers in step 1.

6. A number is *algebraic* if it can be "solved"—that is, if it can be the root of an equation $P(x) = 0$, where $P(x)$ is a polynomial with integer coefficients. Show that every rational number $\frac{p}{q}$, where p and q are integers and $q \neq 0$, is an algebraic number.

7. Show that every number of the form $p + \sqrt{q}$, where p and q are integers, is an algebraic number.

8. Some numbers cannot be "solved"—that is, they cannot be solutions of any equation $P(x) = 0$, where $P(x)$ is a polynomial with integer coefficients. These numbers are called *transcendental*. List some numbers that you think are transcendental. Compare your list with others in your class.

Solving Complex Numbers

Name _____

Solve That Number—Part 2

Complex numbers are often the roots of equations of the form $P(x) = 0$, where $P(x)$ is a polynomial with integer coefficients. Such complex numbers are also algebraic.

Think of "solving" a complex number such as $2 - 3i$ as the process of finding an equation $P(2 - 3i) = 0$, where $P(x)$ is a polynomial with integer coefficients, and $2 - 3i$ is a root of the equation. You can "solve" the complex number $2 - 3i$ by the following method:

> Let $x = 2 - 3i$.
>
> Subtracting 2 from both sides gives you $x - 2 = -3i$.
>
> By squaring both sides, you get $x^2 - 4x + 4 = -9$.
>
> Adding 9 to both sides gives you $x^2 - 4x + 13 = 0$.
>
> Thus, you know that $2 - 3i$ is a root of the equation $x^2 - 4x + 13 = 0$.

1. "Solve" the following complex numbers.

   a. $4 + 7i$ _____

   b. $4 - 7i$ _____

   Assume that p and q are integers and "solve" the following:

   c. $p + qi$ _____

   d. $p - qi$ _____

Navigating through Number and Operations in Grade 9–12

Solving Complex Numbers (continued)

Name _____

2. The complex numbers $a + bi$ and $a - bi$ are *conjugates* of each other. From your results in step 1, make a conjecture about the results of "solving" conjugate complex numbers.

   Explain your thinking.

3. The beginning of the activity sheet showed you a method for "solving" the complex number $2 - 3i$. You could generalize this method for any complex number $p + qi$, as follows:

   $x = p + qi$

   $x - p = qi$

   $(x - p)^2 = x^2 - 2px + p^2 = -q^2$

   $x^2 - 2px + p^2 + q^2 = 0$

   How could you use these steps to prove that any complex number $p + qi$, where p and q are integers, is algebraic?

4. Consider the following general claim:

   If $P(x)$ is a polynomial with integer coefficients, then $P(h + ki)$ and $P(h - ki)$ are conjugates.

   Examine, for example, $P(x) = 2x^3 - 5x + 9$ when x equals $3 - 8i$ or $3 + 8i$:

   - Substituting $3 - 8i$ for x in $P(x) = 2x^3 - 5x + 9$ gives
     $2(3 - 8i)^3 - 5(3 - 8i) + 9 = -1104 + 632i$.

   - Substituting $3 + 8i$ for x gives $2(3 + 8i)^3 - 5(3 + 8i) + 9 = -1104 - 632i$.

   Note that the two results, $-1104 + 632i$ and $-1104 - 632i$, are conjugates of each other.

Solving Complex Numbers (continued)

Name _____

a. Explore the general claim with technology. Use a variety of polynomials for $P(x)$ (including polynomials of different degrees) and choose a variety of complex numbers for $h + ki$. Summarize your findings.

b. Without technology, prove that the general claim is true when $P(x)$ is any quadratic polynomial and $x = 3 + 8i$ and $x = 3 - 8i$. *Hint:* Use the method shown at the beginning of step 4 and evaluate $P(3 + 8i)$ and $P(3 - 8i)$ for $P(x) = ax^2 + bx + c$.

c. If you have access to a computer algebra system (CAS), prove that $P(h + ki)$ is the conjugate of $P(h - ki)$ for all quadratic and cubic polynomials with integer coefficients.

5. a. Assuming that the general claim in step 4 is true, explain how it leads to the conclusion that $h + ki$ is a root of a polynomial with integer coefficients if and only if $h - ki$ is also a root of the polynomial.

b. What does 5(a) tell you about the set of complex numbers that are algebraic?

Frequencies, Scales, and Guitars

Name _____

Consider the following graph, which shows sound-wave data collected with a microphone from a tuning fork that sounds the note A:

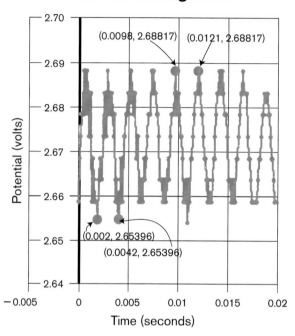

The number of complete waves that the note produces during each second of time is its *frequency*. A frequency of one wave per second is called a *hertz* (Hz). Note that the graph appears to show nine complete waves in the time from 0 to 0.02 seconds.

The time required for one complete wave is called the *period* of the wave. This is the time that elapses between two consecutive peaks or two consecutive troughs. The period is the reciprocal of the frequency.

1. *a.* Use the *x*-values that the graph identifies for two peaks and two troughs to find two different estimates of the period of the wave.

    0.0121 seconds − 0.0098 seconds = _____ seconds
    0.0042 seconds − 0.0020 seconds = _____ seconds

    b. Why are these estimates different?

Note that in the graph the label for the *y*-axis is "Potential (volts)." An air wave causes a membrane in the microphone to move, and that membrane's motion is part of an electrical capacitance system and causes a fluctuation in the electrical potential of the capacitance system. Electrical potential is measured in volts.

Frequencies, Scales, and Guitars (continued)

Name _____

c. Average your two estimates to find a single estimate of the frequency of the tuning fork.

Since ancient times, musicians have known a very important and useful fact: If they play a note on a string that they press down at the midpoint between the instrument's bridges, they will produce a note that is the same as, but one octave higher than, the note that they make on the open, "unpressed" string. As the illustration shows, the twelfth fret on a guitar is halfway between the instrument's bridges, one of which is called the *nut*.

On a guitar and other stringed instruments, including pianos, each octave is divided into twelve notes, a semitone apart.

2. Consider the graphs of sound-wave data (below) collected from a guitar's A-string when it was open (left) and when it was pressed at the 12th fret (right).

   The two notes that the string produces under the different conditions are one octave apart.

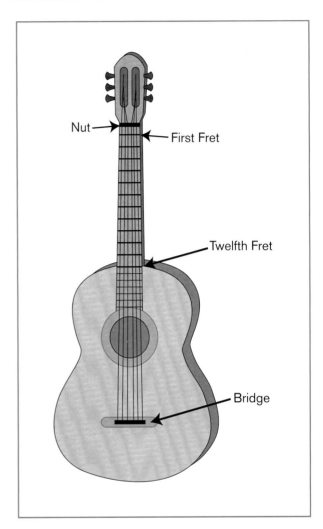

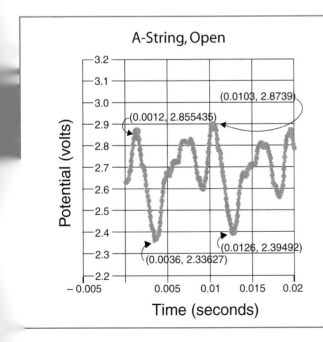

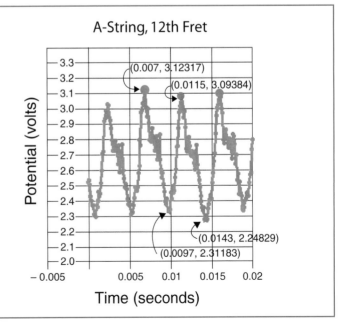

Frequencies, Scales, and Guitars (continued)

Name _____

   a. Complete table 1 to show the period and the frequency of a note produced on an A-string when it is open and when a musician presses it at the 12th fret.

Table 1

	Open String				12th Fret			
	Time 1	Time 2	Period (seconds)	F = 1/P (hertz)	Time 1	Time 2	Period (seconds)	F = 1/P (hertz)
Peaks	0.0012	0.0103			0.0070	0.0115		
Troughs	0.0036	0.0126			0.0097	0.0143		

   b. How does the frequency of the higher (12th fret) note compare with that of the lower note?

3. Table 2 (on p. 109) gives the frequencies of the indicated notes on a guitar.

   a. Complete columns 4 and 5 by computing the successive differences and ratios.

   b. Do you find a pattern in your results? _____ Explain.

4. Refer again to table 2.

   a. Average your values in the ratio column, and compare your average with the solution of the equation $220 = 110n^{12}$.

   b. Find a simple function that gives a good approximation of the frequencies of the notes in table 2 in terms of their fret numbers.

   c. Use your function from part (*b*) to predict the frequencies of the notes in the octave *above* the octave shown in table 2. Record your predictions in table 3 (also on p. 109).

5. Consider table 4 (on p. 110), which gives the distance from the bridge of a guitar to each of the first twelve frets.

   a. Complete columns 3 and 4 by computing the differences and ratios of the consecutive lengths.

   b. Do you find a pattern in your results? _____ Explain.

Frequencies, Scales, and Guitars (continued)

Name _____

Table 2

Fret	Note	Frequency (Hz)	Difference (Hz) $(F_{n+1} - F_n)$	Ratio $\left(\dfrac{F_{n+1}}{F_n}\right)$
0, or "open"	A	110	117 − 110 =	$\dfrac{117}{110} =$
1	A#	117	123 − 117 =	$\dfrac{123}{110} =$
2	B	123		
3	C	131		
4	C#	139		
5	D	147		
6	D#	156		
7	E	165		
8	F	175		
9	F#	185		
10	G	196		
11	G#	208		
12	A	220		

Table 3

Note	Predicted Frequency
A	
A#	
B	
C	
C#	
D	
D#	
E	
F	
F#	
G	
G#	
A	

Navigating through Number and Operations in Grade 9–12

Frequencies, Scales, and Guitars (continued)

Name _____

6. Assume that the consecutive ratios in table 4 are exactly the same and are all equal to the ratio r.

   a. Explain why the distance from the bridge to the 12th fret — that is, 32.4 cm — would have to equal $(64.8)r^{12}$.

   b. Find an exact solution for the equation $32.4 = (64.8)r^{12}$. Is your solution a rational number?

   c. Explain how to use the ratio r to compute the locations of frets on a guitar.

7. Working with your findings from steps 3, 5, and 6, explain how the distances to the different frets relate to the frequencies of the notes that a musician produces by pressing the string at those frets.

Table 4

Fret	Length (cm)	Difference (cm) $(L_{n+1} - L_n)$	Ratio $\left(\dfrac{L_{n+1}}{L_n}\right)$
0 (Nut)	64.8	$61.2 - 64.8 = -3.6$	
1	61.2		
2	57.8		
3	54.6		
4	51.4		
5	48.7		
6	45.9		
7	43.2		
8	40.9		
9	38.5		
10	36.3		
11	34.3		
12	32.4		

Like Clockwork

Name _____

Rock Around the Clock—Part 1

Suppose that you have signed up for Crypto 101, an introductory course in cryptography. The class sounds exciting — cryptographers code and decode messages to ensure the security of personal information in such environments as the Internet. On the first day, your teacher divides the class into groups and distributes a message to decode: *TPOOOJLXFVLYH?F.*

How can you and your classmates decode this message? Would you expect to use mathematics? Would it surprise you to learn that mathematics is at the heart of many of today's encoding and decoding schemes?

One of the primary mathematical tools behind these schemes is modular arithmetic, also known as "clock arithmetic." What is clock arithmetic, and how is it used in coding and decoding messages? Consider the "clock" at the right. It has seven "stops," numbered 0 through 6, inside the main circle:

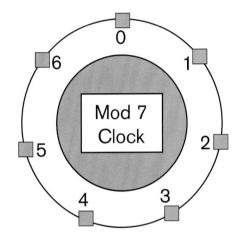

To "walk" a number — for example 20 — around the clock, you would start at 0 and take 20 steps clockwise around the clock. Note that each step places you on one of the numbered stops on the clock. So, for instance, taking 7 steps would bring you back to 0. Taking 11 steps would place you at 4.

1. In the clock below, someone has walked the numbers from 1 to 9, recording the stopping position of each number outside the main circle:

    a. Continue this person's work, walking the numbers from 10 to 20 around the clock and writing each number's stopping position as you go.

    b. Where does 20 stop?

2. Look for patterns in all the numbers that stop in the same spot on the clock.

    a. What patterns do you see?

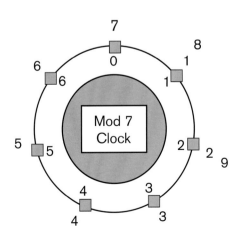

Like Clockwork (continued)

Name _____

    b. Will these patterns continue for higher numbers?

3. Suppose that you walked the number 3925 around the clock.

    a. Where would 3925 stop?

    b. Can you describe a shortcut for determining its stop without "walking"?

    c. Compare strategies within your group.

4. The notation $[3925]_7$ refers to the stopping point for 3925. Find the following stopping points, sharing any patterns that you notice and comparing strategies.

    a. $[39]_7$

    b. $[7 \times 69 + 2]_7$

    c. $[7 \times 439 + 39]_7$

    d. $[7n + 3]_7$

5. Could you use your calculator to find the place where the number 3925 would stop if you walked it around the clock?

    a. Explain your strategy.

    b. Compare your strategy with those of others in your group.

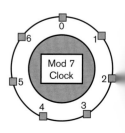

Like Clockwork (continued)

Name _____

6. Find the following stopping points, and look for shortcuts, comparing strategies with those of others in your group.

   a. $[3 \times 45]_7$

   b. $[62 \times 36]_7$

   c. $[12 \times (16 \times 43)]_7$

7. The division algorithm for integers ensures that for any positive integer p, unique integers q and r exist such that $p = 7q + r$, where $0 \le r < 7$. Express the following numbers in the form $7q + r$, where $0 \le r < 7$.

   a. 405 _____

   b. $7 \times 439 + 39$ _____

8. When a number is expressed in the form $7q + r$, where $0 \le r < 7$, what is the value of $[7q + r]_7$?

9. Suppose that n and m are positive integers in the form $7a + b$, where a and b are integers, and $0 \le b < 7$. Let $n = 7q + r$ and $m = 7p + s$. Prove each of the following statements:

   a. $[n \times m]_7 = [n \times [m]_7]_7 = [n \times s]_7$

   Hint: Replace m by $7p + s$.

   b. $[n \times m]_7 = [r \times s]_7$

   Hint: Use part (a) and replace n by $7q + r$.

10. What do your results in step 9 imply about walking a product of two numbers around the clock?

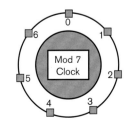

In part 2 of Rock Around the Clock, you will continue to investigate the role of clock arithmetic in coding and decoding messages.

Encryption à la Mod

Name _____

Rock Around the Clock—Part 2

Let's call the integers 1, 2, 3, 4, 5, and 6 the *mod 7 encryption numbers* and define a new mod 7 multiplication operation, denoted by $\otimes$, in the following manner: $a \otimes b = [a \times b]_7$. In other words, to find $a \otimes b$, we would multiply a and b in the regular way and then find the remainder of this product when it is divided by 7.

1. Complete the operation table for the mod 7 system.

    Table 1

$\otimes$	1	2	3	4	5	6
1						
2						
3						
4						
5						
6						

2. Table 1 reveals many properties of the mod 7 system. Refer to the table in answering the following questions:

    a. If c and a are mod 7 encryption numbers, then is it possible for $c \times a$ to be a multiple of 7? _____
    Explain.

    b. Is the operation $\otimes$ closed on the mod 7 encryption numbers? _____ Explain. *Note:* If the operation is *closed*, then when c and a are mod 7 encryption numbers, so is $c \otimes a$.

    c. If c and a are mod 7 encryption numbers, then is it always true that $c \otimes a = a \otimes c$? _____
    Explain whether the operation $\otimes$ is commutative for all mod 7 encryption numbers.

Encryption à la Mod (continued)

Name _____

d. If a, b, and c are mod 7 encryption numbers and $a \neq b$, then is it possible that $c \otimes a = c \otimes b$? _____ Explain.

e. If c is any mod 7 encryption number, then does a unique mod 7 encryption number c^{-1} exist such that $c \otimes c^{-1} = c^{-1} \otimes c = 1$? _____ Explain whether every mod 7 encryption number has an inverse under the operation $\otimes$.

3. Some properties of $\otimes$ are not clear from table 1. For example, is $\otimes$ associative? How could you determine whether or not $a \otimes (b \otimes c) = (a \otimes b) \otimes c$ for any mod 7 encryption numbers a, b, and c?

There are many other systems like the mod 7 system. Some have different properties. Take, for example the mod 6 encryption numbers, consisting of the integers 1, 2, 3, 4, and 5. Define mod 6 multiplication, again denoted by $\otimes$, in the following manner: $a \otimes b = [a \times b]_6$. In other words, to find $a \otimes b$, you would multiply a and b in the regular way and then find the remainder of this product when it is divided by 6.

4. Complete the following operation table for the mod 6 system:

Table 2

$\otimes$	1	2	3	4	5
1					
2					
3					
4					
5					

Encryption à la Mod (continued)

Name _____

5. Refer to table 2 in answering the following questions about the properties of the mod 6 system.

   a. If c and a are mod 6 encryption numbers, then is it possible for $c \times a$ to be a multiple of 6? _____ Explain.

   b. Is the operation $\otimes$ closed on the mod 6 encryption numbers? _____ Explain.

   c. If c and a are mod 6 encryption numbers, then is it always true that $c \otimes a = a \otimes c$? _____ Explain whether the operation $\otimes$ is commutative for all mod 6 encryption numbers.

   d. If a, b, and c are mod 6 encryption numbers and $a \neq b$, then is it possible that $c \otimes a = c \otimes b$? _____ Explain.

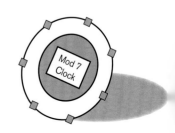

116 Navigating through Number and Operations in Grades 9–12

Encryption à la Mod (continued)

Name _____

e. If c is any mod 6 encryption number, then does a unique mod 6 encryption number c^{-1} exist such that $c \otimes c^{-1} = c^{-1} \otimes c = 1$? _____ Explain whether every mod 6 encryption number has an inverse under the operation $\otimes$.

6. Can you identify a difference between the numbers 6 and 7 that explains the differences between the mod 6 and mod 7 systems?

Modular arithmetic is a useful tool for coding and decoding messages. Consider how the process works for the mod 31 encryption numbers and mod 31 multiplication. Since 31 is a prime number, all the mod 31 encryption numbers have inverses in mod 31 multiplication, as do all the mod 7 encryption numbers in mod 7 multiplication. For example, $2 \otimes 16 = [2 \times 16]_{31} = [32]_{31} = 1$ in the mod 31 system. Thus, 2 and 16 are inverses of each other. Likewise, -2 and -16 multiply to give 32. Since -2 corresponds to 29 on the mod 31 clock and -16 corresponds to 15, then 29 and 15 are also multiplicative inverses.

7. Complete table 3 to show the multiplicative inverse of each of the mod 31 encryption numbers.

Table 3

x	1	2	3	4	5	6	7	8	9	10	11	12	13	14	15
x^{-1}	1	16													29

x	16	17	18	19	20	21	22	23	24	25	26	27	28	29	30
x^{-1}	2													15	

In part 3 of Rock Around the Clock, you will see how cryptographers use the fact that all mod 31 encryption numbers have multiplicative inverses that are unique mod 31 encryption numbers.

Ciphering in Mod 31

Name _____

Rock Around the Clock—Part 3

Cryptographers frequently use a substitution-value table to help them convert the "plaintext" of a message to "ciphertext" so that others cannot easily decipher it. A substitution-value table is a list that pairs numerical values with the alphabetic characters and punctuation marks in a message.

A common cipher technique requires that the number of substitution values be one less than a prime number. Take the prime number 31, for example. Thirty substitution values can represent all twenty-six letters in the English alphabet plus four punctuation marks. Table 1 presents one possible substitution scheme:

Table 1
Substitution Values for Plaintext

A	B	C	D	E	F	G	H	I	J	K	L	M	N	O
1	2	3	4	5	6	7	8	9	10	11	12	13	14	15
P	Q	R	S	T	U	V	W	X	Y	Z	–	.	,	?
16	17	18	19	20	21	22	23	24	25	26	27	27	29	30

Note that the character for the value 27 is "–" (a short dash). The substitution scheme uses this character for the space between words.

Suppose that you wish to encode the message *I have a secret*. Consider table 2, which is similar to one that a cryptographer might use to code or decode such a message.

Table 2
Encoding the Message I have a secret.

Plaintext	I	–	H	A										
Position values	1	2	3	4										
Substitution values	9	27	8	1										
Ciphertext values	9	23	24	4										
Ciphertext	I	W	X	D										

Ciphering in Mod 31 (continued)

Name _____

Follow steps 1–5 to complete the table.

1. Complete row 1 in table 2 by entering the remaining letters and punctuation marks of the message in order, one letter or one punctuation mark to a box. (Remember that the message ends in a period, and be sure to use a dash for the space between words.)

2. Complete row 2 of table 2 by entering the "position value" of each character or punctuation mark (including every dash) in the corresponding box. The first character in the plaintext is assigned a position value of 1, the second character is assigned a position value of 2, and so on.

   (In the substitution scheme that table 1 sets up, position values may range from 1 to 30. If a message is longer than thirty characters, then the counting scheme begins again at 1 with the thirty-first character.)

3. Referring to table 1, complete row 3 of table 2 by entering the "substitution value" of each character or punctuation mark in the corresponding box.

4. Row 4 of table 2 gives the "ciphertext value" of each character or punctuation mark in the message. Modular arithmetic comes into play in the determination of this value. Complete row 4 by using mod 31 multiplication to find the product of each character's position value and its substitution value.

   (For example, to compute the ciphertext value for the second character in the message, "–", multiply, mod 31, 2 by 27. Two is the position value of "–" and 27 is its substitution value from table 1. The result is 23, which then becomes the ciphertext value for the "–" in position 2.)

5. The ciphertext values in row 4 serve as the numerical base for the corresponding encoded ciphertext symbols in row 5. Complete row 5 by using the substitution values in table 1 again, this time to come up with a ciphertext symbol for each ciphertext value.

   (For example, the ciphertext value for "–" in position 2 is 23. Table 1 gives *W* as the substitution for 23, so *W* is the ciphertext symbol in row 5.)

6. When you have encoded the entire message, write your ciphertext here:

In part 4 of Rock Around the Clock, you will apply what you have learned in parts 1–3 and use mod 31 to encode and decode messages.

Make a Code / Break a Code

Name _____

Rock Around the Clock—Part 4

Now you're set to use mod 31 to encode and decode messages. The substitution table that you worked with in part 3 of Rock Around the Clock appears below.

Table 1
Substitution Values for Plaintext

A	B	C	D	E	F	G	H	I	J	K	L	M	N	O
1	2	3	4	5	6	7	8	9	10	11	12	13	14	15
P	Q	R	S	T	U	V	W	X	Y	Z	–	.	,	?
16	17	18	19	20	21	22	23	24	25	26	27	27	29	30

Continue to use this substitution table as you encode new messages and explore the process of decoding a message.

Encoding

1. Complete table 2 to encode the message *Math is useful.*

Table 2
Encoding the Message Math is useful.

Plaintext														
Position values														
Substitution values														
Ciphertext values														
Ciphertext														

Encoded message _____

120   Navigating through Number and Operations in Grades 9–12

Make a Code / Break a Code (continued)

Name _____

2. Complete table 3 to encode your first name (up to 12 characters).

Table 3
Encoding My First Name

Plaintext												
Position values												
Substitution values												
Ciphertext values												
Ciphertext												

Encoded name _____

Decoding

How would you go about *decoding* a message that you received in this code? Decoding becomes relatively simple if you use a fact that you discovered in part 2 of Rock Around the Clock — that all the mod 31 encryption numbers have multiplicative inverses.

The person who encrypted the message encoded each character by multiplying its position value by its substitution value. Thus, to decode the message, you must multiply each ciphertext value by the *multiplicative inverse of its position value*. Table 4 shows the inverses that you found in part 2.

Table 4
Multiplicative Inverses for Mod 31 Encryption Numbers

x	1	2	3	4	5	6	7	8	9	10	11	12	13	14	15
x^{-1}	1	16	21	8	25	26	9	4	7	28	17	13	12	20	29

x	16	17	18	19	20	21	22	23	24	25	26	27	28	29	30
x^{-1}	2	11	19	18	14	3	24	27	22	5	6	23	10	15	30

Make a Code / Break a Code (continued)

Name _____

To see how the decoding process works, look again at the ciphertext of the message *I have a secret*.

IWXDQ?CHZDXEQHUN

Consider the fifth character, *Q*. To decode *Q*, you would multiply 17, which is *Q*'s substitution value, by the multiplicative inverse of 5, which is *Q*'s position value. Under mod 31 multiplication, the multiplicative inverse of 5 is 25.

$$17 \otimes 25 = [17 \times 25]_{31} = 22$$

Referring to table 1, note that 22 corresponds to *V*. Thus, the *Q* in the fifth position decodes to *V*.

The following table shows the values for decoding the rest of the message.

Table 5
Decoding the Message IWXDQ?CHZDXEQHUN

Ciphertext	I	W	X	D	Q	?	C	H	Z	D	X	E	Q	H	U	N
Position values	1	2	3	4	5	6	7	8	9	10	11	12	13	14	15	16
Inverses of position values	1	16	21	8	25	26	9	4	7	28	17	13	12	20	29	2
Substitution values	9	23	24	4	17	30	3	8	26	4	24	5	17	8	21	14
Plaintext values	9	27	8	1	22	5	27	1	27	19	5	3	18	5	20	28
Plaintext	I	–	H	A	V	E	–	A	–	S	E	C	R	E	T	.

3. In part 1 of Rock Around the Clock, you imagined that you were a student in Crypto 101, an introductory course in cryptography. On the first day, your teacher asked you and your classmates to decode the message *TPOOOJLXFVLYH?F*. Use the process that you have just learned to complete table 6 (on the next page) and decode the message.

Make a Code / Break a Code (continued)

Name _____

Table 6
Decoding the Message Presented to the Students in Crypto 101: TPOOOJLXFVLYH?F

Ciphertext														
Position values														
Inverses of position values														
Substitution values														
Plaintext values														
Plaintext														

Decoded message _____

The four parts of Rock Around the Clock have demonstrated the importance of mathematics in cryptography. Could you use mod 31, or another modular system, to encode the message *Math is essential to cryptography*? Do you think a friend in another class would be able to decipher your message? What would your friend need to know?

Probing the Pattern

Name _____

Number Triangles—Part 1

1. The numbers in triangle A have a particular pattern. Find the pattern and use it to figure out the missing numbers *x*, *y*, and *z* in triangle B.

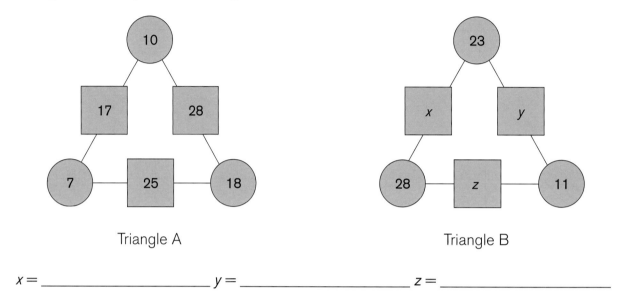

Triangle A   Triangle B

x = _____ *y* = _____ *z* = _____

2. Look again at triangles A and B.

   a. Describe the pattern that you found in step 1.

   b. If you change the numbers in the circles of triangle B to other whole numbers, will your pattern continue to provide solutions for *x*, *y*, and *z*?

   c. What other patterns can you find in the numbers in a triangle such as triangle A?

Probing the Pattern (continued)

Name _____

Let's say that a *number triangle* is a triangle that has integers *a*, *b*, and *c* at its vertices and integers *x*, *y*, and *z* on its sides, as shown in the illustration.

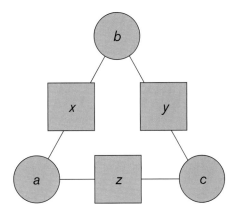

In a number triangle, $x = a + b$, $y = b + c$, and $z = a + c$.

3. Consider integers that are odd.

    a. Can a number triangle have exactly one odd integer? _____

    b. Can a number triangle have exactly two odd integers? _____

    c. What are the possible arrangements of odd integers in a number triangle?

4. Imagine that you can use any integers, including negative integers, as numbers in a number triangle.

    a. Can you make a number triangle that has exactly one negative integer?

    b. Can you make a number triangle that has exactly two negative integers?

    c. What are the possible arrangements of negative integers in a number triangle?

Navigating through Number and Operations in Grade 9–12

Probing the Pattern (continued)

Name _____

5. Complete the following number triangles.

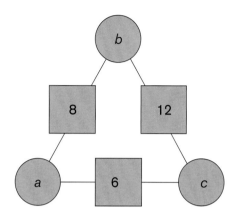 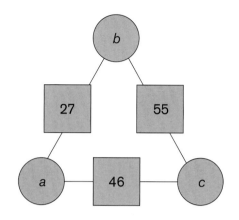

6. Discuss the strategy that you used in step 5 with others in your class.

   a. Did everyone use the same strategy? _____

   b. What strategy do you think is best for doing this kind of problem?

It All Adds Up

Name _____

Number Triangles—Part 2

A *number triangle* is a triangle that has integers *a*, *b*, and *c* at its vertices and integers *x*, *y*, and *z* on its sides, as shown in the illustration.

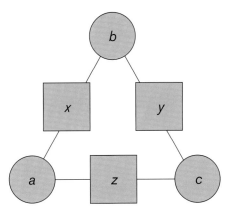

In a number triangle, $x = a + b$, $y = b + c$, and $z = a + c$.

1. Consider number triangles A and B below.

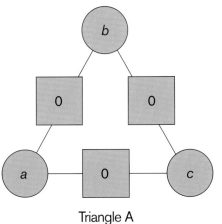
Triangle A

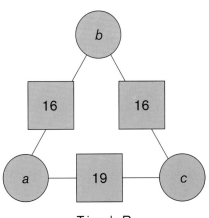
Triangle B

   a. Find all the solutions to the triangles.

   Triangle A _____

   Triangle B _____

   b. If no solution is possible, explain why.

   c. Use your explanation to guide you in finding another triangle for which no solution is possible.

Navigating through Number and Operations in Grade 9–12

It All Adds Up (continued)

Name _____

2. If *x*, *y*, and *z* are positive integers in the following number triangle below, under what conditions will *a*, *b*, or *c* necessarily be negative?

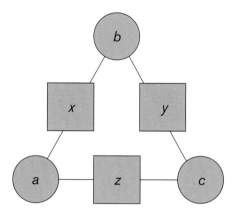

3. Consider a number triangle like that in step 2.

   a. What must be true if one or more zeroes appear as values for *x*, *y*, or *z* in the triangle?

   b. What must be true about *x*, *y*, and *z* if one or more zeroes appear as values for *a*, *b*, or *c*?

4. Suppose that you have a number triangle in which *x*, *y*, and *z* are known and *a*, *b*, and *c* are unknown.

   a. Create several such number triangles and solve them.

   b. Can you create a triangle for which *a*, *b*, and *c* have more than one solution? _____

   c. Justify your answer in 4(*b*).

5. When you have a number triangle like that in step 4, in which *x*, *y*, and *z* are known and *a*, *b*, and *c* are unknown, under what conditions can you find no solution to the triangle?

Take a Trip around a Triangle

Name _____

Number Triangles—Part 3

A *number triangle* is a triangle that has integers *a*, *b*, and *c* at its vertices and integers *x*, *y*, and *z* on its sides, as shown in the illustration at the right.

In a number triangle, $x = a + b$, $y = b + c$, and $z = a + c$.

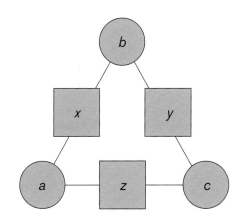

1. Express the relationships among *a*, *b*, and *c* in the triangle at the right as a system of equations.

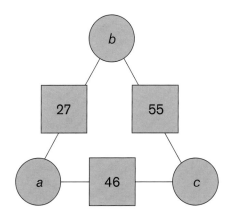

2. Solve the system of equations in step 1 to find values for *a*, *b*, and *c*.

3. In the number triangle at the right, if $x + y + z$ is an odd number, prove that no integers *a*, *b*, and *c* exist that will complete the triangle.

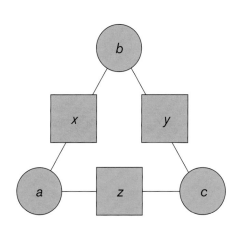

Navigating through Number and Operations in Grade 9–12

Take a Trip around a Triangle (continued)

Name _____

4. Suppose that a student solves the number triangle shown at the right and explains her method as follows:

   "Say that I guess that $a = 4$. Because $b = 8 - a$, I have to say that $b = 4$. I also know that $c = 12 - b$, so c has to be 8. I know too that $a = 6 - c$, which gives me $a = -2$. But I started with $a = 4$. OK. So how about if I split the difference between 4 and -2, or average them? This gives me $a = 1$. If I use this value for a, I force b to be 7 and c to be 5. Look! Now I have a solution that works!"

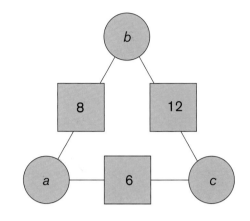

   a. Try to solve the following number triangles with the student's strategy.

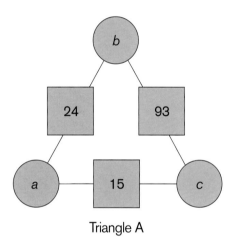

Triangle A

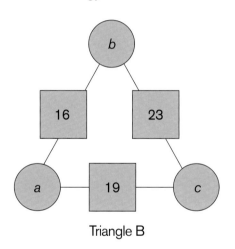

Triangle B

   b. Did the strategy work? _____

5. Continue to examine the strategy from step 4.

   a. Make and solve problems of your own by filling in the squares in the number triangles below with integers.

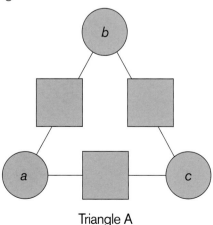

Triangle A

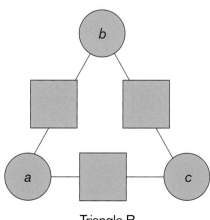

Triangle B

Take a Trip around a Triangle (continued)

Name _____

b. Try to solve your problems with the strategy from step 4.

c. Did the strategy work? _____

6. Assume that a number triangle has a solution.

   a. Using the strategy from step 4, can you always find the correct values for *a*, *b*, and *c* in just two trips around the triangle? _____

   b. Justify your answer.

7. Think abstractly about what happens when you apply the strategy from step 4 to a triangle that has a solution. Use the letter *g* to denote your guess for the value of *a*.

   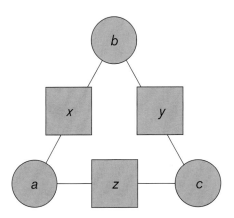

   - You begin by saying that $a = g$.

   - Next, you say that $b = x - g$, and $c = y - b = y - (x - g)$.

   - Your trip around the triangle leads you to a new value for *a*, which you can call a'.

   - You know that $a' = z - c = z - (y - (x - g))$.

   - Averaging *a* and a' gives you $\dfrac{g + (z - (y - (x - g)))}{2}$, or $\dfrac{z - y + x}{2}$.

   Show that this is the correct value for *a*.

Fair and Square

Name _____

Perfect Squares—Part 1

A *perfect square* is a positive integer w such that $w = n^2$ for some positive integer n. Thus, a perfect square has a positive integer as a square root. For example, 25 is a perfect square because $25 = 5^2$. Perfect squares have some properties that other numbers do not.

1. Consider the following numbers:

   121   432 18^4 9^3 $\sqrt{25}$ -4

   a. Which numbers are perfect squares? _____

   b. Justify your answer.

2. Complete the following table for the first ten integers that are perfect squares and the first ten that are not.

First Ten Perfect Squares	Number of Positive Divisors	Prime Factorization in Exponential Form	First Ten Positive Integers That Are Not Perfect Squares	Number of Positive Divisors	Prime Factorization in Exponential Form
1			2		
4			3		
9			5		
16			6		
25			7		
36			8		
49			10		
64			11		
81			12		
100			13		

132   Navigating through Number and Operations in Grades 9–12

Fair and Square (continued)

Name _____

3. Inspect the data on positive divisors in your table.

   a. Make a conjecture about how perfect squares differ from other integers in their numbers of positive divisors.

   b. Give reasons to support your conjecture.

4. Inspect the data on prime factorizations in your table.

   a. Make a conjecture about how perfect squares differ from other integers in their prime factorizations.

   b. Give reasons to support your conjecture.

5. The difference between the 9th and 8th perfect squares in the table is $81 - 64$. This difference equals $9^2 - 8^2$, or 17. Furthermore, $17 = 9 + 8$, and 17 is the 9th odd number counting from 1.

   a. Examine other pairs of perfect squares. Does this pattern hold for your other sample pairs?

   b. If you think that this pattern always holds for the difference between two consecutive perfect squares, use algebra to prove your conjecture.

Navigating through Number and Operations in Grade 9–12   133

Fair and Square (continued)

Name _____

Ancient mathematicians studied perfect squares as square numbers, which they represented in the following geometrical fashion:

6. Use this geometrical representation of perfect squares to show that the pattern always works for the difference between two consecutive perfect squares.

7. Investigate positive integers that are differences between two squares.

    a. Is every odd number a difference of two squares?

    b. Is every even number a difference of two squares?

    c. Justify any claims or conjectures that you make.

Last-Digit Revelations

Name _____

Perfect Squares—Part 2

Perfect squares are often too large for a calculator to handle. For example, the millionth perfect square is 1,000,000,000,000, or $(1,000,000)^2$. Numbers with so many digits are too large for most calculators to compute exactly.

1. Try the perfect square $(23,456,789)^2$ on your calculator.

   a. Is the value that your calculator gives exact, or is it rounded off?

   b. How many digits are in $(23,456,789)^2$? _____ How do you know?

   c. What is the last digit in the number $(23,456,789)^2$? _____ How do you know?

2. Consider the number 2,183,110,734,262.

   a. Is this number a perfect square?

   b. Justify your answer.

3. Two digits are missing in the following number:

   65 _ _ 7.

   a. Is the number a perfect square? _____

   b. Why, or why not?

Navigating through Number and Operations in Grade 9–12 135

Last-Digit Revelations (continued)

Name _____

4. Examine the last digit of each of the first ten perfect squares in the table below.

First Ten Perfect Squares	Last Digit in Square
1	1
4	4
9	9
16	6
25	5
36	6
49	9
64	4
81	1
100	0

  a. Find a pattern and check to see if it continues.

  b. What numbers cannot be the last digit of a perfect square? _____ Why is this so?

5. A *perfect cube* is an integer w such that $w = n^3$ for some positive integer n. Thus, a perfect cube has a positive integer as its cube root. You can study the patterns in the last digits of perfect cubes as you did in the last digits of perfect squares. Again, you can generalize your results to other powers. On the next page, complete the following table of last digits of powers of x for $1 \leq x \leq 10$.

Last-Digit Revelations (continued)

Name _____

x	x^1	x^2	x^3	x^4	x^5	x^6	x^7	x^8
1	1	1	1	1	1	1	1	1
2	2	4	8	6	2	4	8	6
3	3	9						
4	4	6						
5	5	5						
6	6	6						
7	7	9						
8	8	4						
9	9	1						
10	0	0						

6. Examine the data in the table in step 5.

   a. Find a pattern in each row.

   b. Does the pattern continue for higher powers? (Use your calculator to check.)

7. Continue to examine the data in the table in step 5.

   a. Find a pattern in each column.

   b. Does the pattern continue for higher values of x? (Use your calculator to check.)

Navigating through Number and Operations in Grade 9–12

Last-Digit Revelations (continued)

Name _____

8. Consider the number 6,783,409,875,116,892,008,925,467.

   a. Is this number a perfect fourth power? _____ Why, or why not?

   b. Is this number a perfect square? _____ Why, or why not?

9. What is the last digit in 2007^{50}?

The Singles Club

Name _____

Perfect Squares—Part 3

In Math Land, there is a Singles Club with a membership consisting of the numbers 0, 1, 2, 3, 4, 5, 6, 7, 8, and 9 — the single-digit, nonnegative numbers. This exclusive club is closed to outsiders under the club's rules for addition and multiplication.

In Math Land, *closed* means that the sum of any two singles in the club has to be a member of the club. The product of any two singles also has to be a member of the club.

To make sure that the Singles Club stays closed to outsiders, the club's charter requires that its operations be limited to "last-digit addition" and "last-digit multiplication." The club has registered $\oplus$ and $\otimes$ as exclusive symbols for its operations. Computations are simple under these operations. For example, $6 \oplus 7 = 3$ and $5 \oplus 9 = 4$ under last-digit addition, and $6 \otimes 7 = 2$ and $5 \otimes 9 = 5$ under last-digit multiplication.

1. Are the Singles Club's methods of addition and multiplication the same as the regular methods for addition and multiplication of integers? _____ Why, or why not?

2. Consider the operations $\oplus$ and $\otimes$.
   a. Explain how the operations work.

   b. Complete the following operation tables:

   Last-Digit Addition ($\oplus$)

$\oplus$	0	1	2	3	4	5	6	7	8	9
0										
1										
2										
3										
4										
5										
6										
7										
8										
9										

   Last-Digit Multiplication ($\otimes$)

$\otimes$	0	1	2	3	4	5	6	7	8	9
0										
1										
2										
3										
4										
5										
6										
7										
8										
9										

The Singles Club (continued)

Name _____

The Singles Club likes to claim that the operations that guarantee its exclusiveness also offer all the benefits of the usual operations of addition and multiplication on integers. Suppose that the club has hired you as a mathematics consultant to verify this claim. Answer the following questions to help you in this task.

3. Are the operations ⊕ and ⊗ commutative? _____ Explain.

4. Are the operations ⊕ and ⊗ associative? _____ Explain.

5. Does last-digit multiplication distribute over last-digit addition? _____ Explain.

6. Do the operations ⊕ and ⊗ have identity elements that are members of the Singles Club? _____ Explain.

7. In general in Math Land, every integer has an additive inverse, sometimes called its "opposite." For example, the opposite of 4 is –4, since 4 + –4 = 0. Do the members of the Singles Club have opposites in the club under last-digit addition? _____ Explain.

8. In general in Math Land, there are no nonzero integers whose product is 0, and there is no pair of integers except {1, 1} whose product is 1. Do these properties hold for last-digit multiplication? _____ Explain.

9. In Math Land, a *prime number* is an integer $p > 1$ whose only positive divisors are 1 and p. Under this definition and last-digit multiplication, are any members of the Singles Club prime? _____ Explain.

10. Does the Singles Club have any perfect squares under last-digit multiplication? _____ If so, how are they related to the perfect squares under regular multiplication?

11. As you have seen, membership in the Singles Club is strictly limited by the club's unique operations. You have identified some differences between the system of numbers in the Singles Club and the system of integers. List other differences that you notice—for example, are there any perfect cubes in the Singles Club?

Keeping It Legal

Name _____

Flooding a Water World—Part 1

Imagine a world covered with water but with shallow areas that managers of the water world can surround with dikes and pump dry to create habitable land. The following laws regulate the construction of dikes:

> - All dikes must begin at one tower and end at a different one.
> - No dike may join more than two towers.
> - Two dikes may intersect only at a tower.
> - More than one dike may connect two particular towers.
> - All dikes must be connected in a network so that citizens may travel on top of them from any tower to any other tower in the network.

In this world, a canton is a connected region surrounded by a perimeter of dikes and towers to allow the water-world managers to pump it dry. One can travel between any two points in a connected region without passing over a dike or through a tower. A dike is on the perimeter of a region if one of its sides abuts the region and its other side does not.

Figures 1 and 2 show models of networks of dikes and towers that satisfy all the laws of the water world. The models consist of Styrofoam balls that represent the world, pins that represent towers, and rubber bands that represent dikes. The towers and dikes form two different networks.

Fig. 1. A network with three cantons

Fig. 2. A network with no cantons

1. Working with the other members of your group, examine the dikes and towers in the networks pictured in figures 1 and 2.

   a. Verify that the networks in both models satisfy each law of the water world. Discuss any disagreements or questions that you have.

   b. If you have materials such as those in figures 1 and 2, construct your own model of a network that satisfies the laws of the water world. Be creative!

Navigating through Number and Operations in Grade 9–12

Keeping It Legal (continued)

Name _____

2. The three-dimensional models pictured in figures 1 and 2 are equivalent to the two-dimensional *graphs* of the networks shown in figures 3 and 4, respectively.

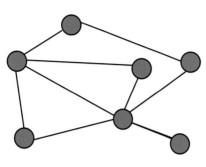 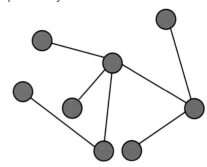

Fig. 3   Fig. 4

  a. Working alone, construct a 3-D model or draw a 2-D graph to show a legal water-world network with many dikes and towers but no cantons. Be creative!

  b. Count and record the numbers of dikes and towers in your network.

  Dikes = _____

  Towers = _____

  c. Compare your results with those of others in your group. What relationship appears to hold between the number of towers and the number of dikes in a network that has no cantons? Be sure that the members of your group agree.

  d. Explain why the relationship holds for any network with no cantons.

3. Working alone, create a legal water-world network with many dikes, towers, and cantons.

  a. Construct a 3-D model or draw a 2-D graph of your network.

142   Navigating through Number and Operations in Grades 9–12

Keeping It Legal (continued)

Name _____

b. Count and record the numbers of dikes and towers in your network.

   Dikes = _____

   Towers = _____

c. Compare your results with those of others in your group. Can you find a relationship between the numbers of towers and dikes in your networks with cantons?

In the water world, the *dike load* of a tower is the number of dikes that intersect at the tower. The managers of the water world claim that the sum of the dike loads in a legal network is twice the number of dikes.

4. Evaluate the managers' claim.

   a. Is the claim true for any network that satisfies the laws of the water world? _____
   Why, or why not?

   b. Verify that your network satisfies the claim.

5. Take another look at your network with cantons, remembering that the laws of the water world require that the network be connected at all times.

   a. If the managers remove a dike from the perimeter of a canton, do they risk disconnecting the network? _____ Why, or why not?

   b. Compare your answer with those of others in your group. Make a conjecture about whether removing one dike from the perimeter of a canton in a legal network can ever disconnect the network.

   c. Prove your conjecture in 5(*b*).

Keeping It Legal (continued)

Name _____

d. When the water-world managers remove a dike from a canton, what is the effect on the total number of dikes, towers, and cantons in a legal network?

Suppose that the managers of the water world have pumped the water from all the cantons in your network and have kept them dry for a number of years. However, they have now decided to flood all your cantons by removing dikes in the network. The following laws govern the removal of dikes to flood cantons:

> • A dike may be removed only if doing so will lead immediately to the flooding of a dry region.
>
> • A dike that has water on both sides (see the protruding dike in fig. 3) may not be removed, since such an action could inadvertently disconnect the towers in a network.

6. Consider the dikes that the managers could remove from your network to flood all your cantons while keeping within the law.

   a. Carefully mark these dikes on your model or drawing. How many dikes would the managers have to remove?

   b. Choose a different set of dikes for the managers to remove. How many dikes would they have to remove with this set?

   c. Compare your answers to 6(*a*) and 6(*b*) with those of others in your group. Make a generalization about the number of dikes that managers must remove to flood all the cantons in a network while complying with water-world law.

   d. Work with the other members of your group on a formula that relates the number of dikes removed to the number of cantons in the original network.

   Number of removed dikes = _____

Keeping It Legal (continued)

Name _____

7. Consider the towers and dikes that remain in your network after the managers have removed dikes and flooded all the cantons.

   a. Double-check to be sure that no cantons remain and that the network is still connected.

   b. What is the relationship between the number of dikes that remain and the number of towers that the network contains?

   c. Compare your answers with those that you gave in steps 2(b) and 2(c). Come up with a formula that relates the number of remaining dikes to the number of towers in the network.

   Number of remaining dikes = _____

8. Consider the formulas that you developed in steps 6(d) and 7(c).

   a. Use these formulas to deduce a new formula that relates the number of dikes in a legal network to the number of towers and cantons in the network.

   Number of dikes = _____

   b. Use the networks in figures 1 and 2 to check your formula.

   c. Explain why your formula holds for all legal networks in the water world.

Other Realms, Other Regions

Name _____

Flooding a Water World—Part 2

Reflect on the work that you did in part 1, "Keeping It Legal." There you created networks of towers and dikes according to the laws of a hypothetical water world. You then removed dikes to flood your cantons — the regions of habitable land that your networks established. Now you will transfer your findings and apply them in other realms.

1. Consider a cube, which is an example of a convex polyhedron. A cube has 6 faces, 8 vertices (or corners), and 12 edges.

    a. Use your formula from step 8(*a*) in part 1 to determine a formula that relates the numbers of faces, edges, and vertices in any convex polyhedron. *Hint*: Think of the polyhedron as a water world with a tower at each vertex and a dike on each edge. Let one face of the polyhedron be water and the other faces be cantons.

    b. Validate your formula by applying it to various three-dimensional solids.

2. Suppose that *n* points are located on a circle so that the chords (or line segments) connecting them satisfy the rule that *no three chords intersect at the same point on the interior*. The following figures show six points located on a circle in ways that satisfy the rule (fig. 1) or break it (fig. 2).

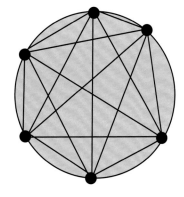

Fig. 1. Six points that fit the rule

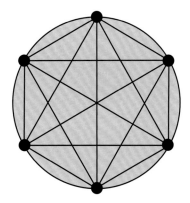

Fig. 2. Six points that do not fit the rule

Other Realms, Other Regions (continued)

Name _____

Can you figure out a formula that tells the number of nonoverlapping regions that the chords create in the circle? In figure 1, the circle has 31 regions — count them. Consider various values for *n* and fill in the following table showing your results. Make sure that the cases you consider satisfy the rule for the locations of the points on the circle!

Number of points (*n*) on circle	1	2	3	4	5	6	7
Number of regions determined						31	

Do you notice any patterns in your table of values? By now, you probably suspect that the formula for the number of regions determined by *n* points on a circle is somewhat complicated! However, to make the formula easier to find, you can use ideas and relationships that you discovered in part 1, "Keeping It Legal":

- You can represent each point where two chords intersect as a *tower*. Figure 3 shows a case in which five points are located on a circle (note that the points where chords intersect on the circle itself — that is, the original five points — are shown as round towers, and the points where chords intersect in the interior of the circle are shown as square towers).

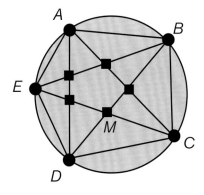

Fig. 3

- You can represent the line segments and the arcs of the circle that connect these points of intersection as *dikes*, just as in the graphs of legal networks in the water world in "Keeping It Legal." In figure 3, there are 25 such segments and arcs, and hence there are 25 dikes. The number of regions that you are trying to count is the same as the number of cantons in the corresponding legal network of towers and dikes.

- You know that the number of cantons in a legal network is given by the following expression:

    Number of cantons = Number of dikes − Number of towers + 1.

Thus, the challenge of coming up with a formula for the number of regions in the circle reduces to finding the numbers of dikes and towers in the corresponding water-world network.

Other Realms, Other Regions (continued)

Name _____

a. Find the number of towers in the network that corresponds to the situation where *n* points are placed on the circle. *Hint:* Note that every choice of four points on the circle determines exactly one point of intersection inside the circle. In figure 3, points *B*, *C*, *D*, and *E* determine a quadrilateral whose diagonals, chords $\overline{BD}$ and $\overline{CE}$, meet at a point *M* inside the circle.

b. Find the number of dikes in the corresponding network by using your results from problem 4 of "Keeping It Legal."

c. Write an expression for the number of nonoverlapping regions that are formed when *n* points are located on a circle.

 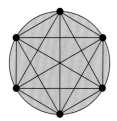 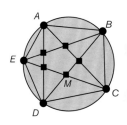

Solutions for the Blackline Masters

Solutions for "When Fractions Are Whole"

1. Students' answers will vary, except in (e), for which the answer given is the only answer. Sample answers follow for (a)–(d):

   a. The fraction $\frac{2}{3}$ is in simplest form and is not equivalent to a positive integer.

   b. The fraction $\frac{3}{1}$ is in simplest form and is equivalent to a positive integer.

   c. The fraction $\frac{6}{2}$ is not in simplest form and is equivalent to a positive integer.

   d. The fraction $\frac{6}{2}$ is not in simplest form, and its square is equivalent to a positive integer.

   e. The fraction $\frac{1}{1}$ is the only fraction that is in simplest form, has a denominator that is a divisor of its numerator, and has a numerator that is a divisor of its denominator.

2. a. Students' answers will vary; $\frac{6}{2}$ is one possible answer.

   b. Not possible; $\frac{p}{1} = p$ for any integer p.

   c. Not possible; if $\frac{p}{q}$ is in simplest form, then p and q have no common prime factors. This means that p^2 and q^2 have no common prime factors, since squaring a number only changes the exponents in its prime factorization, not its prime factors.

   d. Not possible; if $\frac{p}{q} = N$ then $p = N \times q$. This means that q is a divisor of p. Since $q \neq 1$, $\frac{p}{q}$ can be simplified, and this fact contradicts the assumptions.

   e. Not possible; if q is a divisor of p, then $p = N \times q$ for some integer N. Hence, $\frac{p}{q} = N$, a conclusion that contradicts the assumptions.

   f. Not possible; if $\frac{p}{q}$ is not in simplest form, then p and q have a common prime divisor. Hence, p^2 and q^2 have a common prime divisor, a conclusion that contradicts the assumptions.

   g. Not possible; $\frac{p}{1} = p$, an integer. Therefore, $\frac{p^2}{1^2} = \frac{p^2}{1} = p^2$, which is also an integer. This result contradicts the assumptions.

3. a. **Fact 1:** If a fraction is in simplest form, then the square of the fraction _is in simplest form._
   b. **Fact 2:** If a fraction is in simplest form and is equivalent to a positive integer, then its denominator _is 1._

4. a. $\left(\frac{p}{q}\right)^2 = N$.

   b. Yes, by fact 1, which the students established in 3(a), $\frac{p^2}{q^2}$ is in simplest form.

   c. The students can conclude that the denominator q^2 equals 1 by fact 2, which they established in 3(b).

   d. The students can conclude that $N = p^2$. Therefore, $N = 1, 4, 9, 16, 25, \ldots$.

5. When N is a positive integer and $\sqrt{N} = \frac{p}{q}$, a rational fraction in simplest form, then N _is a perfect square_.

6. The students can conclude that if a positive integer N is not a perfect square, then $\sqrt{N} \neq \frac{p}{q}$, a rational fraction in simplest form.

Navigating through Number and Operations in Grades 9–12

7. The number 2 is not a perfect square. Therefore, by the conclusion in step 6, the students can say that $\sqrt{2}$ is irrational.

Solutions for "Designing a Line"

1. Since all the students are using the same unit, their number lines should look approximately the same. The text discusses possible geometric methods for obtaining the specified lengths.

2. *a.* To find $\frac{1+\sqrt{5}}{2}$ palms, a student can start at $\frac{1}{2}$ palms on his or her own number line and then use a group member's line to obtain a measure of $\frac{\sqrt{5}}{2}$ palms to add on to it. The student can then mark the new point on his or her number line.
   b. To find $\pi - \sqrt{2}$ palms, students can use the same procedure as in 2(*a*) but subtract the second length instead of adding it.
   c. As discussed in the text, to find $\pi \times \sqrt{2}$ palms, students can roll a sheet of paper (or construction paper or tagboard) to form a cylinder with a diameter of $\sqrt{2}$ palms. Then they need only measure its circumference.
   d. As discussed in the text, to find $\frac{\pi}{\sqrt{2}}$ palms, students can make a cylinder with a diameter of $\frac{\sqrt{2}}{2}$ palms and measure its circumference.

3. In figure 1, $V = 2 \times 1\frac{1}{2} = 3$ palms, since $\frac{V}{2} = \frac{1\frac{1}{2}}{1}$, by similar triangles. In figure 2, $W = \frac{2\frac{1}{2}}{1\frac{3}{4}} = \frac{10}{7}$ palms, since $\frac{2\frac{1}{2}}{1\frac{3}{4}} = \frac{W}{1}$, by similar triangles.

4. *a.* No, the students' results in step 3 do not depend on the fact that the students joined their number lines to form a right angle.
   b. Other angles would also work since the results depend only on the triangles being similar.

5. In figure 3, $V = AB$, since $\frac{V}{A} = \frac{B}{1}$, by similar triangles. In figure 4, $W = \frac{B}{A}$, since $\frac{B}{A} = \frac{W}{1}$, by similar triangles.

6. *a.* When the students interchange the values $\sqrt{2}$ and π on the vertical and horizontal axes in the construction, they will find that the result is the same.
   b. This example suggests that "geometric" multiplication is commutative.
   c. The students must show that 1 is the multiplicative identity by showing that $1 \times A = A \times 1 = A$. In step 5, they determined that $V = AB$ in the construction in figure 3. If they replace B by 1, then they will force V to equal A by similar triangles, and they will have $\frac{A}{1} = \frac{A}{1}$, or $1 \times A = A \times 1 = A$.
   d. Students need to show that $\frac{1}{A}$ is the multiplicative inverse for every point A by showing that $A \cdot \frac{1}{A} = 1$. In step 5, they determined that $W = \frac{B}{A}$ in the construction in figure 4. If they replace B by 1, then W is forced to equal $\frac{1}{A}$ by similar triangles, and then they will have $\frac{1}{A} = \frac{\frac{1}{A}}{1}$, or $A \cdot \frac{1}{A} = 1$.

7. The students' answers will vary. For example, students might argue that the product is equal to the limit of the sequence $3 \times 1, 3.1 \times 1.4, 3.14 \times 1.41, 3.141 \times 1.414, \ldots$.

8. *a.* Responses will vary, but students may want to show one or more cubits on their number lines, using the fact that the Rhind Papyrus treats a palm as 1/7 of a cubit.
   b. Responses will vary. Some students might suggest that including negative numbers would be unrealistic even though they could produce these numbers with the construction methods and operations in the activity. Other students might question the authenticity of including such numbers as the complex number *i*, or the base of the natural logarithm, *e*. In addition, students could investigate the history of mathematics in an effort to date

the discoveries of irrational numbers and the Pythagorean theorem and place these innovations in relation to the ancient Egyptian measurement system with cubits and palms from 2000 B.C.

Solutions for "Trigonometric Target Practice"

1. a. In terms of h and k, $\tan \theta = \frac{k}{h}$.

   b. The equation of the line is $y = \tan(\theta) \cdot x$.

   c. The slope of the line is $\tan \theta$.

2. a. The slope must be $\frac{q}{p}$.

   b. Students' examples will vary. In general, a line l that passes through the origin and lattice point (p, q) will also pass through any lattice point (r, s) such that $\frac{s}{r} = \frac{kq}{kp} = \frac{q}{p}$. If $\frac{s}{r} = \frac{q}{p}$, then $s = \frac{q}{p} \cdot r$, which means that (r, s) satisfies the equation of the line $y = \frac{q}{p} \cdot x$.

3. a. The graph does not pass through any lattice point other than $(0, 0)$.

   b. No line that the students graph will appear to pass through a lattice point other than $(0, 0)$. (See the discussion on pages 24–25.)

   c. A good approximation for $\tan \theta$ is $\frac{q}{p}$, since the slope of the line through $(0, 0)$ and (p, q) would have to be very close to the slope of $y = \tan(\theta) \cdot x$.

4. Students should conclude that $\tan \theta \neq \frac{q}{p}$ for any integers p and q. They should conclude, in other words, that $\tan \theta$ is irrational.

Solutions for "Counting Primes"

1. a. Make sure that your students also label the negative multiples of 5.

   b. No matter what two multiples of 5 the students choose, the difference will be a multiple of 5.

   c. No, if the difference between two numbers on the number line is 1, both numbers cannot be multiples of 5. Multiples of 5 are always at least five units apart on the number line.

   d. If $h = 5n$ and $k = 5m$ are two multiples of 5, then $h - k = 5n - 5m = 5(n - m)$.

2. a–c. Students' numbers will vary, but all students should find that if the difference between two numbers is 1, both cannot be multiples of their number.

   d. Students should also find, and be able to prove, that if n is an integer, then the difference between two multiples of n is also a multiple of n. Furthermore, they should see that multiples of n are at least n units apart on the number line. Thus, no two multiples of n can differ by 1 unless $n = 1$.

3. No, no positive integers other than 1 can be divisors of both M and $M + 1$. If $d > 0$ is a divisor of both M and $M + 1$, then their difference, which is 1, would have to be a multiple of d. This fact forces d to equal 1.

4. Students should circle 2, 3, 5, 7, 11, 13, 17, 19, 23, 29, and 31. Be sure that they *do not* circle 1, 0, or any negative integers.

5. *a–c.* Students' numbers and factorizations will vary.
6. *a.* Some of the students' numbers may be prime, and others will be composite.

    b. Students' answers about the prime factors of *N* will vary. However, no prime factors of *N* will be on the list of six primes that the student originally chose.

    c. The students will obtain a remainder of 1 when they divide their number *N* by any of their selected primes.

    d. No matter what set of primes a student chooses, the results will be the same.

7. *a.* The integers $p_1 \times p_2 \times p_3 \times p_4 \times p_5 \times p_6$ and $p_1 \times p_2 \times p_3 \times p_4 \times p_5 \times p_6 + 1$ are consecutive on the number line, and hence they are relatively prime.

    b. Regardless of the size of the finite set *S*, the number that is the product of all of its elements and the number that is 1 more than that product will always be relatively prime since they are consecutive integers.

    c. To find another prime that is not in the set *S*, the students could find the prime factorization of $p_1 \times p_2 \times \ldots \times p_n + 1$. None of its prime factors would be in *S*, since the prime numbers in *S* are the same as the prime factors of $p_1 \times p_2 \times \ldots \times p_n$. But, $p_1 \times p_2 \times \ldots \times p_n$ and $p_1 \times p_2 \times \ldots \times p_n + 1$ are consecutive integers and therefore have no prime factors in common. (Note that $p_1 \times p_2 \times \ldots \times p_n + 1$ might itself be a prime number.)

8. The set of all prime numbers must be infinite, since no matter how many primes are in the set, there are always primes that are not in it. If $S = \{p_1, p_2, p_3, \ldots, p_n\}$ is the set of primes, then there are always two possibilities. Either $(p_1 \times p_2 \times p_3 \times \ldots \times p_{n+1}) + 1$ is itself prime, or it has prime factors that are not in *S*. Both cases yield new primes for set *S*.

Solutions for "Adding Complex Numbers"

Complex Numbers and Matrices—Part 1

1. *a.* $(3 + 4i) + (6 + 3i) = 9 + 7i.$

    b. $(2 - 4i) + (-3 + i) = -1 - 3i.$

    c. $(-5 - 2i) + (4 - 3i) = -1 - 5i.$

2. *a.* The completed table is shown:

(x_1, y_1)	(x_2, y_2)	$(x_1 + x_2, y_1 + y_2)$
(3, 4)	(6, 3)	(9, 7)
(2, –4)	(–3, 1)	(–1, –3)
(–5, –2)	(4, –3)	(–1, –5)

    b. Students' examples and explanations will vary but should establish the following correspondence:

    $(a + bi) + (c + di) = (a + c) + (b + d)i$
    $\updownarrow \quad \updownarrow \quad \updownarrow \quad \updownarrow$
    $(a, b) + (c, d) = (a + c, \; b + d)$

3. *a.* The origin, the endpoints of the two vectors being added, and the endpoint of the resultant vector determine a parallelogram.

b. The graphs show the parallelograms for the two sums.

$(2 - 4i) + (-3 + i) = (-1 - 3i)$

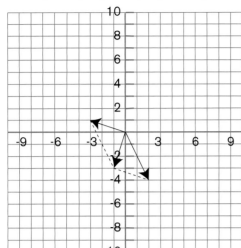

$(-5 - 2i) + (4 - 3i) = (-1 - 5i)$

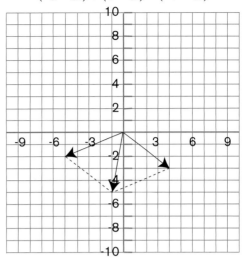

c. Students' examples will vary.

4. a. $\begin{bmatrix} 3 & 4 \\ -4 & 3 \end{bmatrix} + \begin{bmatrix} 6 & 3 \\ -3 & 6 \end{bmatrix} = \begin{bmatrix} 9 & 7 \\ -7 & 9 \end{bmatrix}$

b. $\begin{bmatrix} 2 & -4 \\ 4 & 2 \end{bmatrix} + \begin{bmatrix} -3 & 1 \\ -1 & -3 \end{bmatrix} = \begin{bmatrix} -1 & -3 \\ 3 & -1 \end{bmatrix}$

c. $\begin{bmatrix} -5 & -2 \\ 2 & -5 \end{bmatrix} + \begin{bmatrix} 4 & -3 \\ 3 & 4 \end{bmatrix} = \begin{bmatrix} -1 & -5 \\ 5 & -1 \end{bmatrix}$

5. Yes. Students' examples and explanations will vary but should establish the following correspondence:

$(a + bi) + (c + di) = (a + c) + (b + d)i$

$\begin{bmatrix} a & b \\ -b & a \end{bmatrix} + \begin{bmatrix} c & d \\ -d & c \end{bmatrix} = \begin{bmatrix} a+c & b+d \\ -(b+d) & a+c \end{bmatrix}$

Solutions for "Multiplying Complex Numbers"

Complex Numbers and Matrices—Part 2

1. a. $(3 + 4i)(6 + 3i) = 6 + 33i.$
   b. $(2 - 4i)(-3 + i) = -2 + 14i.$
   c. $(-5 - 2i)(4 - 3i) = -26 + 7i.$

2. a. $\begin{bmatrix} 3 & 4 \\ -4 & 3 \end{bmatrix} \begin{bmatrix} 6 & 3 \\ -3 & 6 \end{bmatrix} = \begin{bmatrix} 6 & 33 \\ -33 & 6 \end{bmatrix}.$

b. $\begin{bmatrix} 2 & -4 \\ -4 & 2 \end{bmatrix} \begin{bmatrix} -3 & 1 \\ -1 & 3 \end{bmatrix} = \begin{bmatrix} -2 & 14 \\ -14 & -2 \end{bmatrix}.$

c. $\begin{bmatrix} -5 & -2 \\ 2 & -5 \end{bmatrix} \begin{bmatrix} 4 & -3 \\ 3 & 4 \end{bmatrix} = \begin{bmatrix} -26 & 7 \\ -7 & -26 \end{bmatrix}.$

3. a. Yes, multiplying two matrices of the form $\begin{bmatrix} a & b \\ -b & a \end{bmatrix}$ always yields a matrix of the same form.

b. Students' examples and explanations will vary but should establish the following correspondence:

$\begin{bmatrix} a & b \\ -b & a \end{bmatrix} \times \begin{bmatrix} c & d \\ -d & c \end{bmatrix} = \begin{bmatrix} ac - bd & ad + bc \\ -(ad + bc) & ac - bd \end{bmatrix}$

$\updownarrow \qquad\qquad \updownarrow \qquad\qquad\qquad \updownarrow$

$(a + bi) \times (c + di) = (ac - bd) + (ad + bc)i$

4. No, matrix multiplication is not commutative in general. In the example,

$A \times B = \begin{bmatrix} -8 & 26 \\ -4 & -8 \end{bmatrix}$ and $B \times A = \begin{bmatrix} 2 & -17 \\ 12 & -18 \end{bmatrix}.$

5. Students should find that the three examples commute. In general, multiplication is commutative on the set of matrices of the form $\begin{bmatrix} a & b \\ -b & a \end{bmatrix}.$

6. a and b. Solving $(3 + 4i)(a + bi) = (1 + 0i)$ and $\begin{bmatrix} 3 & 4 \\ -4 & 3 \end{bmatrix} \begin{bmatrix} a & b \\ -b & a \end{bmatrix} = \begin{bmatrix} 1 & 0 \\ 0 & 1 \end{bmatrix}$ for a and b, we get

$a = \dfrac{3}{25}$ and $b = \dfrac{-4}{25}$. Therefore, the inverses are, respectively, $\dfrac{3}{25} - \dfrac{4}{25}i$ and $\begin{bmatrix} \dfrac{3}{25} & \dfrac{-4}{25} \\ \dfrac{4}{25} & \dfrac{3}{25} \end{bmatrix}$. They correspond to each other just as $a + bi$ corresponds to $\begin{bmatrix} a & b \\ -b & a \end{bmatrix}.$

c. Multiplying the matrices in 6(b) and setting components equal yields the same system of equations as in 6(a). Therefore, the results are the same.

7. Students' responses will vary. Students might make a comparison of the properties of the two systems. For example, they might note that both addition and multiplication are commutative. Other students might say that the systems are basically the same, observing that the elements in one exactly match up with the elements of the other, and when you operate on these elements with addition or multiplication, the sums or products match up.

Solutions for "Solving Real Numbers"

Solve That Number—Part 1

1. Students' responses will vary. The following equations are examples of possible "solutions" for the given numbers:
   a. $x - 5 = 0$.
   b. $x + 3 = 0$.
   c. $7x - 2 = 0$.
   d. $x^3 - 6 = 0$.
   e. $x^2 - 2x - 11 = 0$.

2. Yes, one method will work for all the numbers. The easiest method for "solving" the number in 1(e), $\sqrt{12}+1$, is a method that the students can use to solve all the numbers. Using this method, they would let $x = \sqrt{12}+1$. Hence, $x - 1 = \sqrt{12}$ and $(x-1)^2 = 12$. Therefore, $x^2 - 2x - 11 = 0$. So $\sqrt{12}+1$ is a root of $x^2 - 2x - 11 = 0$.

3–5. For each number r in step 1, if r is a root of $P(x) = 0$, where $P(x)$ is a polynomial with integer coefficients, then the students can generate other equations with different polynomials by multiplying $P(x)$ by a constant.

6. Students can show that every rational number $\frac{p}{q}$, where p and q are integers and $q \neq 0$, is algebraic, since $\frac{p}{q}$ is a root of $qx - p = 0$.

7. Students can show that every number $p + \sqrt{q}$, where p and q are integers, is algebraic by first letting $x = p + \sqrt{q}$. Then they can subtract, getting $x - p = \sqrt{q}$, and square both sides, obtaining $(x - p)^2 = q$. Expanding will then give them $x^2 - 2px + p^2 - q = 0$. Thus, they will show that $p + \sqrt{q}$ is a root of $x^2 - 2px + p^2 - q = 0$.

8. The numbers π, $\pi + 4$, e, $e + 1$, and $\log(2)$ are a few samples of transcendental numbers. Many numbers that the students deal with are transcendental. For example, if $x \neq 0$ is a rational number of radians, then $\sin(x)$, $\cos(x)$, and $\tan(x)$ are transcendental numbers. If $x \neq 0$ is a rational number, then $\ln(x)$ and e^x are transcendental numbers. Furthermore, the sum of an algebraic number and a transcendental number is transcendental.

Solutions for "Solving Complex Numbers"

Solve That Number—Part 2

1. Students' responses will vary. The following equations are examples of possible "solutions" for the given numbers:
   a. $x^2 - 8x + 65 = 0$.
   b. $x^2 - 8x + 65 = 0$.
   c. $x^2 - 2px + p^2 + q^2 = 0$.
   d. $x^2 - 2px + p^2 + q^2 = 0$.

2. Students should conjecture that if a complex number z is a root of $P(x)$, a polynomial with real coefficients, then the conjugate of z is also a root of $P(x)$.

3. Using the generalized method for "solving" $p + qi$, the students will obtain $x^2 - 2px + p^2 + q^2 = 0$. They can use this result to prove that any complex number $p + qi$, where p and q are integers, is algebraic, since the coefficient $-2p$ and $p^2 + q^2$ are also integers. That is, the resulting equation is a polynomial with integer coefficients, as required.

4. a. The students' work with technology should help them verify the general claim that $P(a + bi)$ is the conjugate of $P(a - bi)$ when $P(x)$ is a polynomial with integer coefficients.
   b. To prove without technology that the general claim is true for $x = 3 + 8i$ and $x = 3 - 8i$ when $P(x)$ is any quadratic polynomial $ax^2 + bx + c$, students can substitute into $P(x) = ax^2 + bx + c$. Thus, $P(3 + 8i) =$

$a(3 + 8i)^2 + b(3 + 8i) + c = (-55a + 3b + c) + (48a + 8b)i$, and $P(3 - 8i) = a(3 - 8i)^2 + b(3 - 8i) + c = (-55a + 3b + c) - (48a + 8b)i$. These expressions are conjugates of each other.

   c. Students' responses will vary.

5. a. Assuming that the claim in step 4 is true, the students can say that if $P(h + ki) = 0$, then $P(h - ki)$ is the conjugate of 0, since $P(h - ki)$ is the conjugate of $P(h + ki)$. But 0 is its own conjugate. Thus, $P(h - ki) = 0$.
   b. Since $h + ki$ is a root of a polynomial if and only if $h - ki$ is also a root of the polynomial, we can say that if a complex number is algebraic, then so is its conjugate.

Solutions for "Frequency, Scales, and Guitars"

1. a. The students should find 0.0023 seconds and 0.0022 seconds as estimates of the period ($0.0121 - 0.0098 = 0.0023$ seconds; $0.0042 - 0.0020 = 0.0022$ seconds).
   b. The estimates are different because the ordered pairs used for peaks and troughs are only approximations.
   c. Averaging the two estimates of the period as 0.00225 seconds, the students can calculate the frequency (f) as follows: $f \approx \dfrac{1 \text{ wave}}{0.00225 \text{ second}} \approx 444 \text{ waves per second} = 444 \text{ Hz}$.

2. a. The completed table follows:

Table 1

	Open String				12th Fret			
	Time 1	Time 2	Period (seconds)	F = 1/P (hertz)	Time 1	Time 2	Period (seconds)	F = 1/P (hertz)
Peaks	0.0012	0.0103	0.0091	110	0.0070	0.0115	0.0045	222
Troughs	0.0036	0.0126	0.0900	111	0.0097	0.0143	0.0046	217

   b. The frequency of the higher (12th fret) note is about double that of the note on the open string.

3. a. The completed table is shown. (If your students gather their own data, they may use them instead of the data given in the problem.)

Table 2

Fret	Note	Frequency (Hz)	Difference (Hz) $(F_{n+1} - F_1)$	Ratio $\left(\dfrac{F_{n+1}}{F_n}\right)$
0	A	110	7	1.06
1	A#	117	6	1.05
2	B	123	8	1.07
3	C	131	8	1.06
4	C#	139	8	1.06
5	D	147	9	1.06
6	D#	156	9	1.06
7	E	165	10	1.06
8	F	175	10	1.06
9	F#	185	11	1.06
10	G	196	12	1.06
11	G#	208	12	1.06
12	A	220		

   b. The students should find that the ratios in column 5 are approximately constant.

4. *a.* The ratio that the students obtain from the table is approximately 1.06, and when they solve the equation $220 = 110n^{12}$, they find that n is equal to $\sqrt[12]{2}$, which is also approximately 1.06.

  b. A simple function that gives a good approximation of the frequency of a note in terms of its fret number (n) is $Frequency(n) = 110 \times (\sqrt[12]{2})^n$.

  c. The completed table is shown. The frequencies are essentially double those in table 2, and they are very close to the standard frequencies.

Table 3

Note	Frequency (Hz)
A	220
A#	233
B	245
C	261
C#	277
D	294
D#	311
E	330
F	349
F#	370
G	392
G#	415
A	440

5. *a.* The completed table follows:

Table 4

Fret	Length (cm)	Difference (cm) $(L_{n+1} - L_n)$	Ratio $\left(\dfrac{L_{n+1}}{L_n}\right)$
0	64.8	−3.6	0.94
1	61.2	−3.4	0.94
2	57.8	−3.2	0.94
3	54.6	−3.2	0.94
4	51.4	−2.7	0.95
5	48.7	−2.8	0.94
6	45.9	−2.7	0.94
7	43.2	−2.3	0.95
8	40.9	−2.4	0.94
9	38.5	−2.2	0.94
10	36.3	−2	0.94
11	34.3	−1.9	0.94
12	32.4		

  b. The students will find that the ratios in column 4 are approximately constant.

6. *a.* If the consecutive ratios are a constant ratio r, then $\dfrac{61.2}{64.8} = r$, or $61.2 = 64.8r$. Likewise, $57.8 = 61.2r$.

  But this implies that $57.8 = 64.8r^2$. Continuing in this fashion, we can see that $32.4 = 64.8r^{12}$. In general, if we assume that the consecutive ratios are a constant ratio r, then the distance (or length of the guitar string) from the primary bridge to the nth fret is given by the function $Length(n) = 64.8r^n$.

b. If $32.4 = 64.8r^{12}$, then $r = \dfrac{1}{\left(\sqrt[12]{2}\right)} \approx 0.94$. No, the solution is not a rational number.

c. Using r to compute the locations of frets on a guitar, we know that the nth fret will be a distance of $64.8r^n$ cm from the primary bridge.

7. Students' answers will vary but should note that the value for the ratio of consecutive lengths in 6(b) (≈ 0.94) is the reciprocal of the value for the ratio of consecutive frequencies in 4(b) (≈ 1.06). In general, the shorter the distance, the higher the frequency. Students might conjecture that the frequency of the note at fret n is inversely proportional to the length of the string from the primary bridge to fret n. They can prove this by combining the equations $L(n) = 64.8 \times \left(\dfrac{1}{\sqrt[12]{2}}\right)^n$ and $Frequency(n) = 110 \times (\sqrt[12]{2})^n$, where n = the fret number:

$$Frequency(n) = 110 \times \dfrac{64.8}{L(n)} = \dfrac{7128}{L(n)}.$$

Solutions for "Like Clockwork"

Rock Around the Clock—Part 1
1. a. The diagram shows the stopping positions of the numbers.

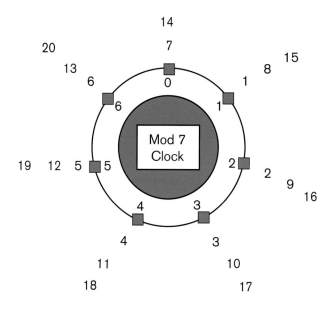

   b. The number 20 stops at 6.
2. a. Answers will vary, but the students should notice that all the numbers at each stop have the same remainder when divided by 7 and differ from one another by multiples of 7.
   b. Yes, these patterns will continue for higher numbers.
3. a. The number 3295 would stop at 5.
   b. Students should discover that as a shortcut they can divide the number by 7: $3925 = 7(560) + 5$. Therefore, 3925 must stop at 5 since $7(560)$, or any multiple of 7, leaves one at 0 on the clock.
   c. Students' strategies may vary.
4. a. $[39]_7 = 4$.
   b. $[7 \times 69 + 2]_7 = 2$.
   c. $[7 \times 439 + 39]_7 = 4$.
   d. $[7n + 3]_7 = 3$.

Students should notice that they can cast out multiples of 7 before computing the remainders. For instance, in (c), $[7 \times 439 + 39]_7 = [39]_7$.

5. a. The strategies that the students use may vary, depending on the calculators that they have. Students using a TI-73 programmable calculator, for example, will enter **Remainder(3925,7)**. Encourage your students to suggest alternate methods.

   b. Students' answers may vary.

6. a. $[3 \times 45]_7 = 2$.

   b. $[62 \times 36]_7 = 6$.

   c. $[12 \times (16 \times 43)]_7 = 3$.

   Students should discover that as a shortcut they can find the remainder for each factor, multiply these remainders, and then find the remainder of the product. For example, to find $[62 \times 36]_7$, they can note that $[62]_7 = 6$ and $[36]_7 = 1$. Therefore, $[62 \times 36]_7 = [6 \times 1]_7 = 6$.

7. a. $405 = 7(57) + 6$.

   b. $7 \times 439 + 39 = 7(444) + 4$.

8. $[7q + r]_7 = r$.

9. a. Substituting for m gives

$$[n \times m]_7 = [n \times (7p + s)]_7 = [7np + ns]_7 = [n \times s]_7.$$

   b. Using the information gained in 9(a) and substituting for n gives

$$[n \times m]_7 = [n \times s]_7 = [(7q + r) \times s]_7 = [7qs + rs]_7 = [r \times s]_7.$$

10. The students should see that their results from step 9 imply that the value of $[n \times m]_7$ does not change if n, m, or both are replaced by their remainders modulo 7.

Solutions for "Encryption à la Mod"

Rock Around the Clock—Part 2

1. The completed operation table for mod 7 follows:

Table 1

⊗	1	2	3	4	5	6
1	1	2	3	4	5	6
2	2	4	6	1	3	5
3	3	6	2	5	1	4
4	4	1	5	2	6	3
5	5	3	4	6	4	2
6	6	5	1	3	2	1

Navigating through Number and Operations in Grades 9–12

2. a. No, it is not possible for $c \times a$ to be a multiple of 7. Otherwise, the remainder of $(c \times a) \div 7$ would be 0. The students should note that no remainder in the table is 0.
   b. Yes, the operation ⊗ is closed on the mod 7 encryption numbers. This should be evident to the students by looking at the numbers in the table. By the division algorithm for integers, the remainder of a division by 7 must be 0, 1, 2, 3, 4, 5, or 6. But in 2(a) they just ruled out a remainder of 0. Thus, they know that the operation is closed on the mod 7 encryption numbers.
   c. Yes, the operation ⊗ is commutative. Since $c \times a = a \times c$, $[c \times a]_7 = [a \times c]_7$. The symmetry of the numbers in the table with respect to the main diagonal shows the students that $c \otimes a = a \otimes c$.
   d. No, it is not possible for $c \otimes a = c \otimes b$. This should be evident to the students from the operation table. No row or column in the table contains the same number twice. The students should note that each row and each column contain each of the numbers 1, 2, 3, 4, 5, and 6 exactly once.
   e. Yes, a unique mod 7 encryption number c^{-1} exists such that $c \otimes c^{-1} = c^{-1} \otimes c$. This should be evident from the fact that each row and each column of the operation table contain the number 1 exactly once.
3. Students should conclude that the operation ⊗ is associative, but their reasons for thinking so may vary. Some students may argue that since in every case they try, $a \otimes (b \otimes c) = (a \otimes b) \otimes c$, ⊗ is associative. Students can also use the results of their work in step 9 in part 1("Like Clockwork") to prove that ⊗ is associative. The discussion of part 2 in the text provides the proof (see pp. 54–55).
4. The completed operation table for mod 6 follows:

Table 2

⊗	1	2	3	4	5
1	1	2	3	4	5
2	2	4	0	2	4
3	3	0	3	0	3
4	4	2	0	4	2
5	5	4	3	2	1

5. a. Yes, it is possible for $c \times a$ to be a multiple of 6; $3 \times 2 = 6$, and hence, $[3 \times 2]_6 = 0$. Likewise $4 \otimes 3 = 0$.
   b. No, the operation ⊗ is not closed on the mod 6 encryption numbers. This should be evident to the students from 5(a); $[3 \times 2]_6 = 0$, which is not a mod 6 encryption number.
   c. Yes, the operation ⊗ is commutative.
   d. Yes, it is possible for $c \otimes a = c \otimes b$. The students will note, for example, that $2 \otimes 2 = 2 \otimes 5 = 4$ in mod 6 multiplication.
   e. No, some mod 6 encryption numbers do not have an inverse for ⊗. That is, there is not always a unique mod 6 encryption number c^{-1} such that $c \otimes c^{-1} = c^{-1} \otimes c = 1$. The numbers 2, 3, and 4 do not have inverses.
6. The students should note that 7 is prime, but 6 is not. Moreover, they should see that all the mod 7 encryption numbers are relatively prime to 7, but the mod 6 encryption numbers 2, 3, and 4 are not relatively prime to 6. Only the elements 1 and 5 are relatively prime to 6 and behave like the mod 7 encryption numbers, all of which have multiplicative inverses.
7. The completed table shows the multiplicative inverse of each of the mod 31 encryption numbers:

Table 3

x	1	2	3	4	5	6	7	8	9	10	11	12	13	14	15
x^{-1}	1	16	21	8	25	26	9	4	7	28	17	13	12	20	29
x	16	17	18	19	20	21	22	23	24	25	26	27	28	29	30
x^{-1}	2	11	19	18	14	3	24	27	22	5	6	23	10	15	30

Solutions for "Ciphering in Mod 31"

Rock Around the Clock—Part 3

1–5. The completed table shows the creation of a ciphertext for the message *I have a secret.*

Table 2
Encoding the message I have a secret.

Plaintext	I	–	H	A	V	E	–	A	–	S	E	C	R	E	T	.
Position values	1	2	3	4	5	6	7	8	9	10	11	12	13	14	15	16
Substitution values	9	27	8	1	22	5	27	1	27	19	5	3	18	5	20	28
Ciphertext values	9	23	24	4	17	30	3	8	26	4	24	5	17	8	21	14
Ciphertext	I	W	X	D	Q	?	C	H	Z	D	X	E	Q	H	U	N

6. As shown in table 2, the ciphertext is *IWXDQ?CHZDXEQHUN.*

Solutions for "Make a Code / Break a Code"

Rock Around the Clock—Part 4

1. The completed table follows:

Table 2
Encoding the message Math is useful.

Plaintext	M	A	T	H	–	I	S	–	U	S	E	F	U	L	.
Position values	1	2	3	4	5	6	7	8	9	10	11	12	13	14	15
Substitution values	13	1	20	8	27	9	19	27	21	19	5	6	21	12	28
Ciphertext values	13	2	29	1	11	23	9	30	3	4	24	10	25	13	17
Ciphertext	M	B	,	A	K	W	I	?	C	D	X	J	Y	M	Q

2. Students' tables for encoding their names will vary.

Navigating through Number and Operations in Grades 9–12

3. The completed table follows:

Table 6
Decoding the Message Presented to the Students in Crypto 101: TPOOOJLXFVLYH?F

Ciphertext	T	P	O	O	O	J	L	X	F	V	L	Y	H	?	F
Position values	1	2	3	4	5	6	7	8	9	10	11	12	13	14	15
Inverses of position values	1	16	21	8	25	26	9	4	7	28	17	13	12	20	29
Substitution values	20	16	15	15	15	10	12	24	6	22	12	25	8	30	6
Plaintext values	20	8	5	27	3	12	15	3	11	27	18	15	3	11	19
Plaintext	T	H	E	–	C	L	O	C	K	–	R	O	C	K	S

Solutions for "Probing the Pattern"

Number Triangles—Part 1
1. The missing values in triangle B are $x = 51$, $y = 34$, and $z = 39$.
2. *a.* The students should notice that the number in each square is the sum of the two numbers in the adjacent circles.
   b. Yes, if the students change the numbers in the circles to other whole numbers, the pattern will continue to provide solutions.
   c. The students' observations about other patterns will vary. Some students will realize that they can get 17 and 28 in the two squares of triangle A if they start with 7 and 18 — the numbers in the two circles below them — and add 10 to both of them. Thus, 17 and 28 must differ by the same amount that 7 and 18 do. The difference between two "consecutive" squares is the same as the difference between two "consecutive" circles of the triangle. Some students might also notice that the sum of the numbers in the squares is equal to twice the sum of the numbers in the circles.
3. *a.* No, a number triangle cannot have exactly one odd number.
   b. No, a number triangle cannot have exactly two odd integers.
   c. Number triangles offer essentially four possibilities for even and odd numbers. These arrangements depend on how many even and odd numbers are at the vertices of the triangle. A number triangle can have 0, 1, 2, or 3 odd numbers at the vertices. The possibilities follow:

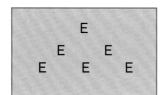 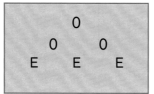 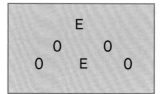 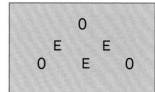

4. *a.* Yes, the students can make a number triangle that has exactly one negative number.
   b. Yes, the students can make a number triangle that has exactly two negative numbers.
   c. The students should conclude that any number of elements in a number triangle can be negative. They should prove this statement with examples and make some observations. For example, if a number triangle has exactly

one negative number, then it cannot be in a square. Likewise, if a number triangle has exactly one positive number, then it cannot be in a square.

5. The students should find that in triangle A, $a = 1$, $b = 7$, and $c = 5$, and in triangle B, $a = 9$, $b = 18$, and $c = 37$.
6. a. Your students will probably report that they used several different strategies.
   b. The students' ideas about the "best" strategy will probably vary. One effective strategy involves applying an observation from the solution to step 2 — that the difference between two "consecutive" squares is the same as the difference between two "consecutive" circles of the triangle. If the students apply this idea to triangle A in step 5, they will see that $c - a = 12 - 8 = 4$. But they also know that $c + a = 6$. Therefore, they can see that $2c = 10$, or $c = 5$.

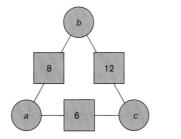
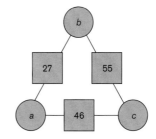

Solutions for "It All Adds Up"

Number Triangles–Part 2

1. a. In triangle A, the students should find that $a = 0$, $b = 0$, and $c = 0$. They should realize that no other solution is possible. They should recognize that because $a + b = a + c = 0$, it must be true that $c = b$. But since $c + b = 0$, the students know that $2c = 2b = 0$. Thus, they know that b and c are 0, forcing a to be equal to 0 as well. The students should find that triangle B has no solution.
   b. To explain why triangle B has no solution, your students might observe that $a + b = b + c = 16$. Therefore, they can say that $a = c$, but they also know that $a + c = 19$. Thus, $2a = 19$, or $a = 9.5$. But this is not a solution, because the solution set for number triangles consists only of integers. Or they might observe that $16 + 16 + 19 = 51$, an odd number. But they should already know that $x + y + z$ is always even because $x + y + z = 2(a + b + c)$.
   c. Students can use their explanation in 1(b) to find many other triangles for which no solution is possible simply by placing an odd number in one of the triangle's squares. (If they completed the activity sheet "Probing the Pattern" and investigated the possible arrangements of odd and even numbers in a number triangle, they will already know that placing an odd number in exactly one of the squares creates an arrangement that has no solution.)

2. The students should note that if $x < b$ or $x < a$, then b or a is negative. If $y < b$ or $y < c$, then b or c is negative. If $z < a$ or $z < c$, then a or c is negative. When x, y, and z are positive integers, at most only one of the values a, b, or c can be negative.

3. a. If in a number triangle, $x = 0$, then $a = -b$. If $x = 0$ and $y = 0$, then $a = -b$ and $a = c$ and $z = 2a$.
   b. If $a = 0$, then the number triangle must have the following form:

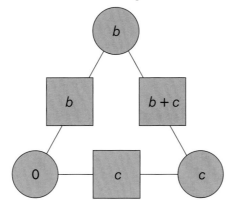

Note that when $a = 0$, $x = b$, $y = b + c$, and $z = c$.

If $a = 0$ and $b = 0$, then the triangle must have the following form:

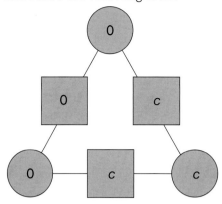

Note that $x = 0$, and y and z are both equal to c.

4. a. Students will create different number triangles in which x, y, and z are known and a, b, and c are unknown.
   b. The students should discover that each such triangle has either no solution or one solution.
   c. Students' justifications for their responses in 4(b) will vary.
5. Students should note that when x, y, and z are known and a, b, and c are unknown, a solution exists if $x + y + z$ is even. If $x + y + z$ is odd, then the number triangle has no solution.

Solutions for "Take a Trip around a Triangle"

Number Triangles—Part 3

1. Students should express the relationships among a, b, and c with the following system of equations:
   $a + b = 27$
   $a + c = 46$
   $b + c = 55$
2. The students can of course solve the system of equations in a variety of ways. Since $a + b = 27$ and $a + c = 46$, they can subtract the equations to get $b - c = -19$. Then they can add the equations $b - c = -19$ and $b + c = 55$ to see that $2b = 36$, or $b = 18$. Substituting, they find that $c = 37$ and $a = 9$.
3. The students can note that $a + b = x$, $a + c = z$, and $b + c = y$. Adding the three equations gives them $2(a + b + c) = x + y + z$. Thus, they know that $x + y + z$ must be even for a solution to exist.
4. a. Make sure that your students use the strategy correctly in the sample triangles.
   b. The students should find that the strategy works perfectly. (See the solutions to steps 6 and 7 below for an explanation of how and why it works.)
5. a. The students' problems will vary.
   b. The numbers that the students place in the three squares must add up to an even number for the triangle to have integer solutions for a, b, and c.
   c. Yes, if the triangle has a solution, the strategy works.
6. a. Yes, the students can always find the values for a, b, and c in just two trips around the triangle.
   b. The students' justifications will vary. They may reason in a manner like the following: "Let a_1 be my first guess. Then $b_1 = x - a_1$, and $c_1 = y - b_1 = y - x + a_1$.
   Therefore, $a_2 = ((z - c_1) + a_1)/2 = (z - y + x - a_1 + a_1)/2 = (z + x - y)/2$.
   Hence, $b_2 = x - a_2 = (x + y - z)/2$ and $c_2 = (y + z - x)/2$.
   None of these values depends on my initial guess."
7. If your students set up the system of equations and solve it, their results should be the same as the values for a_2, b_2, and c_2 in step 6. For the system of equations
   $a + b = x$
   $b + c = y$
   $a + c = z$,

the solution is $a = \frac{z + x - y}{2}$, $b = \frac{x + y - z}{2}$, and $c = \frac{y + z - x}{2}$. Thus, if $x + y + z$ is even, then a, b, and c are integers, and hence there is a unique solution to the number triangle. If $x + y + z$ is odd, then a, b, and c are not integers, and hence there is no solution within the constraints of the problem.

Solutions for "Fair and Square"

Perfect Squares—Part 1

1. *a.* Three of the numbers are perfect squares: 121, 18^4, and 9^3.
   b. $121 = 11^2$, $18^4 = (18^2)^2$, and $9^3 = (3^2)^3 = (3^3)^2$. The other three numbers, 432, $\sqrt{25}$, and -4, are not perfect squares. The students may note that $432 = (2^4)(3^3) = (12^2)(3)$; $\sqrt{25} = \left(\sqrt{5}\right)^2$, but $\sqrt{5}$ is not an integer; and $-4 = (2^2)(-1) = (2i)^2$, but $2i$ is not an integer.
2. The completed table appears in the text as table 4.1 (see p. 68).
3. *a.* Students' conjectures may vary.
   b. Students' reasons may vary, but the students should recognize that perfect squares have odd numbers of divisors since the divisors of N come in pairs unless one of them is $\sqrt{N}$.
4. *a.* Students conjectures may vary.
   b. Students' reasons may vary, but the students should recognize that all the exponents in the prime factorization of a perfect square are even. If $n = m^2$, then the prime factorization of n is the same as the prime factorization of m except that the exponents in n's factorization are double those in m's.
5. *a.* The students should discover that the pattern works only for consecutive perfect squares.
   b. To use algebra to prove that this pattern always holds, students can examine the difference between the two consecutive perfect squares n^2 and $(n-1)^2$: $n^2 - (n-1)^2 = 2n - 1$. The students can readily see that $2n - 1$ is the nth odd number, counting from 1.
6. Using the geometrical representation of perfect squares, students can show that the pattern always works for the difference between two consecutive perfect squares by some method similar to the following:

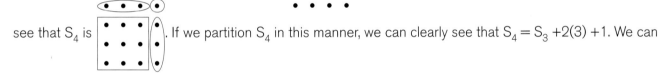

Let S_n be the nth square number. Hence, $S_4 = $ [array shown]. If we look at the array in the right way, we can see that S_4 is [partitioned array shown]. If we partition S_4 in this manner, we can clearly see that $S_4 = S_3 + 2(3) + 1$. We can also see from the diagram that the pattern generalizes to all square numbers. Thus $S_n = S_{n-1} + 2(n-1) + 1$. Therefore, $S_n - S_{n-1} = 2n - 1$.

7. *a.* Every odd integer can be expressed as the difference of two squares.
   b. Every even integer that is a multiple of 4 can be expressed as a difference of two squares.
   c. Solutions are discussed in the text (see p. 69).

Solutions for "Last-Digit Revelations"

Perfect Squares—Part 2

1. *a.* When students try to get the square of $23{,}456{,}789$ on their calculators, they will probably get something like $5.502209502\ E14$. The calculator will have rounded off the value.
   b. There are fifteen digits in the square of $23{,}456{,}789$. This is evident from the scientific notation given by the calculator.
   c. The last digit of $(23{,}456{,}789)^2$ is 1, the last digit in the product 9×9. You might want to explain to your students that the last digit of any positive integer is the remainder when the number is divided by 10. In the

case at hand, $23{,}456{,}789^2 = (23{,}456{,}780 + 9)^2 = 23{,}456{,}780^2 + 18(23{,}456{,}780) + 80 + 1$. The sum of the first three terms is a multiple of 10. Thus, the last digit is 1.

2. *a.* No, the number 2,183,110,734,262 is not a perfect square.
   b. Students can deduce this fact by looking at the last digit. No integer squared equals 2. Perfect squares can end only in 0, 1, 4, 9, 6, or 5.
3. *a.* No, the number 65_ _ 7 cannot be a perfect square.
   b. Students can reason as in 2(*b*). No integer squared equals 7. The students can also use inequalities to come to the same conclusion. They can say that $254 < \sqrt{65007} \le \sqrt{65\_\_7} \le \sqrt{65997} < 257$. Therefore, 65_ _ 7 would have to be either 255^2 or 256^2 if it is to be a perfect square. But it is neither.
4. *a.* The pattern 1, 4, 9, 6, 5, 6, 9, 4, 1, 0 appears in the last digits of the first ten perfect squares. The students should find that the pattern continues.
   b. The last digit of a perfect square cannot be 2, 3, 7, or 8. In the multiplication of two numbers, the product of the last digits determines the last digit in the product. Listing all the squares of single digit numbers reveals that no perfect square can end in 2, 3, 7, or 8.
5. The completed table follows for last digits of powers of x for $1 \le x \le 10$

x	x^1	x^2	x^3	x^4	x^5	x^6	x^7	x^8
1	1	1	1	1	1	1	1	1
2	2	4	8	6	2	4	8	6
3	3	9	7	1	3	9	7	1
4	4	6	4	6	4	6	4	6
5	5	5	5	5	5	5	5	5
6	6	6	6	6	6	6	6	6
7	7	9	3	1	7	9	3	1
8	8	4	2	6	8	4	2	6
9	9	1	9	1	9	1	9	1
10	0	0	0	0	0	0	0	0

6. *a.* The students should notice that every row has a repeating block.
   b. The students should conclude that the pattern continues for higher powers.
7. *a.* The students should notice that every column has a repeating block of 10 numbers.
   b. The students should conclude that the pattern continues for higher values of x.
8. *a.* No, the number 6,783,409,875,116,892,008,925,467 is not a perfect fourth power. A perfect fourth power is also a perfect square, and no perfect square ends in 7.
   b. No, the number is not a perfect square. The solution to 8(*a*) explains why not.
9. The students should observe that the pattern for powers of 7 repeats after every fourth power. Thus, they can deduce that 2007^{48} ends in a 1 and 2007^{50} ends in a 9.

Solutions for "The Singles Club"

Perfect Squares—Part 3
1. The students should observe that the Singles Club's methods are different from those of ordinary addition and multiplication of integers. Ordinary addition and multiplication are not closed on the set {0, 1, 2, 3, 4, 5, 6, 7, 8, 9}.
2. *a.* The operations $\oplus$ and $\otimes$ work like ordinary addition and multiplication of integers except that only the last digit of an ordinary sum or a product is reported as a sum or a product.

b. The completed tables follow:

Last-Digit Addition (⊕)

⊕	0	1	2	3	4	5	6	7	8	9
0	0	1	2	3	4	5	6	7	8	9
1	1	2	3	4	5	6	7	8	9	0
2	2	3	4	5	6	7	8	9	0	1
3	3	4	5	6	7	8	9	0	1	2
4	4	5	6	7	8	9	0	1	2	3
5	5	6	7	8	9	0	1	2	3	4
6	6	7	8	9	0	1	2	3	4	5
7	7	8	9	0	1	2	3	4	5	6
8	8	9	0	1	2	3	4	5	6	7
9	9	0	1	2	3	4	5	6	7	8

Last-Digit Multiplication (⊗)

⊗	0	1	2	3	4	5	6	7	8	9
0	0	0	0	0	0	0	0	0	0	0
1	0	1	2	3	4	5	6	7	8	9
2	0	2	4	6	8	0	2	4	6	8
3	0	3	6	9	2	5	8	1	4	7
4	0	4	8	2	6	0	4	8	2	6
5	0	5	0	5	0	5	0	5	0	5
6	0	6	2	8	4	0	6	2	8	4
7	0	7	4	1	8	5	2	9	6	3
8	0	8	6	4	2	0	8	6	4	2
9	0	9	8	7	6	5	4	3	2	1

3. The operations ⊕ and ⊗ are both commutative. The symmetry in the tables along the main diagonal is a visual indicator of commutativity, but since $a + b = b + a$ and $ab = ba$ under ordinary addition and multiplication of integers, the last digits in these sums and products must be the same under the operations ⊕ and ⊗ as well.
4. The operations ⊕ and ⊗ are both associative. The activity Rock Around the Clock establishes the associative property for modular arithmetic for any base.
5. Yes, last-digit multiplication distributes over last-digit addition. Students know this from the fact that the last digit of an addition under ⊕ is the same as the last digit in the sum of the original numbers under ordinary addition. The same thing is true of the last digit in a product of two numbers.
6. Yes, the operations ⊕ and ⊗ have identity elements that are members of the Singles Club — 0 for last-digit addition and 1 for last-digit multiplication.
7. Yes, the members of the Singles Club have opposites in the club under last-digit addition. In the Single Club, the additive inverse of a number *n* is $10 - n$. Thus, 1 and 9, 2 and 8, 3 and 7, 4 and 6, and 5 and 5 are additive inverses.
8. No, it is not true for Singles Club multiplication that there are no nonzero integers whose product is 0. Students should note that $2 \otimes 5 = 0$, for example. It is also not true for Singles Club multiplication that there is no pair of integers except {1, 1} whose product is 1. Students should note that $7 \otimes 3 = 1$, for example. Students can also inspect the table to discover that these properties do not hold.
9. No, no member of the Singles Club is a prime. Every number n in the Singles Club has other divisors in addition to 1 and *n*.
10. Yes, the Singles Club has the following perfect squares: 0, 1, 4, 5, 6, and 9. This set is obviously different from the set of perfect squares for the integers *x* such that $0 \leq x \leq 9$: {0, 1, 4, 9}. For example, 5 is not a perfect square under ordinary integer multiplication.
11. The students' answers will vary. Students may notice, or example, that although the operation ⊗ has an identity, some numbers have multiplicative inverses under ⊗. This is generally not true of the integers. Students may notice that every member of the singles club is a perfect cube. This is not true of every integer.

Solutions for "Keeping It Legal"

Flooding a Water World—Part 1

1. *a* and *b*. Students' responses will vary.
2. *a* and *b*. Students' responses will vary.
   c. The students should observe that the number of towers is one more than the number of dikes.

d. The students can prove that this relationship holds for any network without cantons by noting that every connected network can start with a single tower to which someone then adds dikes. But in a network with no cantons, any new dike must have one end that attaches to an existing tower while its other end attaches to a new tower—one that was not previously part of the network. (If someone added a new dike to a legal network without cantons by connecting it to two towers already in existence, then it would form a canton.) Thus, every time a legal network that has no cantons gets a new dike, it also gets a new tower. So the number of towers must be one more than the number of dikes.
3. *a* and *b.* Students' responses will vary.
   c. There are no obvious relationships between the number of towers and the number of dikes.
4. *a.* Yes, the managers' claim that the sum of the dike loads in a legal network is twice the number of dikes is true. The laws of the water world require that every dike be connected to exactly two different towers. Therefore, every dike is a member of two different dike loads, and as a result we count each one twice when we sum the dike loads.
   b. Students should find that their networks support the managers' claim.
5. *a–c.* The students should note that removing a dike from the perimeter of a canton does not change the legal status of either of the towers that were the dike's terminal points. These towers remain connected in the network. A canton comes into being when a system of dikes and towers encloses a region. In such a system, there are always two disjoint paths connecting any two towers on the perimeter of the canton. Since there is thus always a detour, or alternate route, for any path that previously traversed a now-removed dike, all the towers in the network must remain connected.
   d. The number of dikes goes down by 1, the number of cantons goes down by 1, and the number of towers remains the same.
6. *a–d.* Students' responses will vary but should lead them to the formula *Number of removed dikes = Number of cantons.* The students should see that number of dikes that water-world managers have to remove is always equal to the number of cantons, no matter what set of dikes the managers remove. There is a one-to-one correspondence between removed dikes and flooded cantons according to the rule that a dike can be removed only if doing so will lead immediately to the flooding of a dry region.
7. *a–c.* In step 2(*c*), the students should have observed that the number of towers in a network without cantons is one more that the number of dikes. They should have proved this result in 2(*d*). Thus, after water-world managers remove dikes to flood all the cantons in a network, the relationship between the number of remaining dikes and the number of towers is the relationship that the students discovered in step 2 for a legal network without cantons. This relationship is given by the formula

   Number of remaining dikes = Number of towers – 1.

8. *a* and *b.* By adding the equation from step 6,

   Number of removed dikes = Number of cantons,

   and the equation from step 7,

   Number of remaining dikes = Number of towers – 1,

   the students can arrive at a new formula that relates the number of dikes in a legal network to the number of towers and cantons that the network contains:

   Number of dikes = Number of removed dikes + Number of remaining dikes

   = Number of cantons + Number of towers – 1.

   c. Since the formulas that the students found in steps 6 and 7 are true for all legal networks that are flooded legally, then the formula in step 8, which the students deduce from these formulas by using basic algebra, is also true for all legal networks.

Solutions for "Other Realms, Other Regions"

Flooding a Water World—Part 2

1. *a* and *b*. Suppose that the polyhedron has F faces, V vertices, and E edges. The students can think of the surface of the polyhedron as a water world, completing the analogy by designating one of the faces as water. They can then think of the remaining $F - 1$ faces as cantons in a water-world network, with dikes (edges) that intersect at towers (vertices) according to the laws of the water world. This reasoning will lead them to the formula

$$E = (F - 1) + V - 1, \text{ or}$$
$$E = F + V - 2.$$

2. The students should complete the table as follows:

Number of points (n) on circle	1	2	3	4	5	6	7
Number of regions determined	1	2	4	8	16	31	57

   a. Twenty-five line segments connect all the points in the example. These line segments divide the circle into 16 regions.

   b. By taking the hint in the problem, students can see the system of line segments, circle segments, and intersection points as a water-world network as long as they count as dikes circle segments and line segments that touch exactly two points of intersection and no more than two line segments intersect at any point in the interior of the circle. Thus, the number of regions in the circle plus the number of intersection points is one more than the number of line segments. But the intersection points include the n original points on the circle and the $\binom{n}{4}$ points of intersection inside the circle. The number of segments connecting intersection points is half the degree sum of the intersection points, according to the claim that the students verified in step 4 of part 1 — that the sum of the dike loads in a legal network is twice the number of dikes. Since each of the n points on the circle joins $n + 1$ segments and each of the intersection points in the circle joins four segments, then the number of segments is $\dfrac{n(n+1) + 4 \times \binom{n}{4}}{2}$. Simplifying and solving for the number of regions, we get

$$R = \binom{n}{4} + \frac{n^2}{2} - \frac{n}{2} + 1 = \frac{n(n-1)(n-2)(n-3)}{4 \times 3 \times 2 \times 1} + \frac{n^2}{2} - \frac{n}{2} + 1.$$

   Therefore, $R = \dfrac{n^4}{24} - \dfrac{n^3}{4} + \dfrac{23n^2}{24} - \dfrac{3n}{4} + 1.$

References

Beckmann, Petr. *A History of π*. New York: St. Martin's Press, 1971.

Beiler, Albert H. *Recreations in the Theory of Numbers*. 2nd ed. New York: Dover Publications, 1966.

Bibby, Neil. "Tuning and Temperament: Closing the Spiral." Chap. 1 in *Music and Mathematics: From Pythagoras to Fractals*, edited by John Fauvel, Raymond Flood, and Robin Wilson, pp. 13–27. New York: Oxford University Press, 2003.

Bonsangue, Martin V., and Gerald E. Gannon. "From Exploration to Generalization: An Introduction to Necessary and Sufficient Conditions." *Mathematics Teacher* 96 (May 2003): 366–71.

Boyer, Carl. *A History of Mathematics*. New York: John Wiley & Sons, 1968.

Bright, George W., Patricia Lamphere Jordan, Carol Malloy, and Tad Watanabe. *Navigating through Measurement in Grades 6–8. Principles and Standards for School Mathematics* Navigations Series. Reston, Va.: National Council of Teachers of Mathematics, 2005.

Burke, Maurice J., and Diana L. Taggart. "So That's Why 22/7 Is Used for π!" Mathematics Teacher 95 (March 2002): 164–69.

Cavanaugh, Mary, Linda Dacey, Carol R. Findell, Carole E. Greenes, Linda Jensen Sheffield, and Marian Small. *Navigating through Number and Operations in Prekindergarten–Grade 2. Principles and Standards for School Mathematics* Navigations Series. Reston, Va.: National Council of Teachers of Mathematics, 2004.

Conway, John H., and Richard K. Guy. *The Book of Numbers*. New York: Springer-Verlag, 1996.

Duncan, Natalie N., Charles Geer, DeAnn Huinker, Larry Leutzinger, Ed Rathmell, and Charles Thompson. *Navigating through Number and Operations in Grades 3–5. Principles and Standards for School Mathematics* Navigations Series. Reston, Va.: National Council of Teachers of Mathematics, 2006.

Francis, Richard L. "Star Numbers and Constellations." *Mathematics Teacher* 86 (January 1993): 88–89.

Hall, Matthew. "Calculator Cryptography." *Mathematics Teacher* 96 (March 2003): 210–12.

Houser, Don. "Roots in Music." *Mathematics Teacher* 95 (January 2002): 16–17.

Johnston, Ian. *Measured Tones: The Interplay of Physics and Music*. 2nd ed. Philadelphia: Institute of Physics Publishing, 2002.

Kalman, Dan. "Six Ways to Sum a Series." *The College Mathematics Journal* 24 (November 1993): 402–21.

Martin, George E. *Transformational Geometry: An Introduction to Symmetry*. New York: Springer-Verlag 1982.

Miller, William A. "Recursion and the Central Polygonal Numbers." Mathematics Teacher 84 (December 1991): 738–46.

National Council of Teachers of Mathematics (NCTM). *Principles and Standards for School Mathematics*. Reston, Va.: NCTM, 2000.

Niven, Ivan. *Numbers: Rational and Irrational*. New York: The Mathematical Association of America, New Mathematical Library, 1964.

Nord, Gail, Eric J. Malm, and John Nord. "Counting Pizzas: A Discovery Lesson Using Combinatorics." *Mathematics Teacher* 95 (January 2002): 8–14.

Parker, Dennis, "Partitioning the Interior of a Circle with Chords." *Mathematics Teacher* 99 (September 2005: 120–24.

Rachlin, Sid, Kathleen Cramer, Connie Finseth, Linda Cooper Foreman, Dorothy Geary, Seth Leavitt, and Margaret Schwan Smith. *Navigating through Number and Operations in Grades 6–8*. Principles and Standards for School Mathematics Navigations Series. Reston, Va.: National Council of Teachers of Mathematics, 2006.

Shiflett, Ray C., and Harris S. Shultz. "An Odd Sum." *Mathematics Teacher* 95 (March 2002): 206–209.

St. John, Dennis. "Exploring Hill Ciphers with Graphing Calculators." *Mathematics Teacher* 91 (March 1998): 240–45.

Stewart, Ian. "Faggot's Fretful Fiasco." Chap. 4 in *Music and Mathematics: From Pythagoras to Fractals*, edited by John Fauvel, Raymond Flood, and Robin Wilson, pp. 61–76. New York: Oxford University Press, 2003.